MODERN SNIPERS

OSPREY
PUBLISHING

DEDICATION

Dedicated to the memory of Gunnery Sergeant Carlos Hathcock, USMC, the father of modern sniping, and to Sergeant Matthew Abbate, USMC, killed in action December 10, 2010, Helmand Province, southern Afghanistan.

MODERN SNIPERS

LEIGH NEVILLE

First published in Great Britain in 2016 by Osprey Publishing
PO Box 883, Oxford, OX1 9PL, UK
1385 Broadway, 5th Floor, New York, NY 10018, USA

E-mail: info@ospreypublishing.com
Osprey Publishing, part of Bloomsbury Publishing Plc
© 2016 Leigh Neville

A CIP catalogue record for this book is available from the British Library
Leigh Neville has asserted his right under the Copyright, Designs and Patents
Act, 1988, to be identified as the Author of this Work.

ISBN: 978 1 4728 1534 7
PDF ISBN: 978 1 4728 1535 4
ePub ISBN: 978 1 4728 1536 1

Index by Zoe Ross
Typeset in Akzidenz-Grotesk and Bembo
Originated by PDQ Media, Bungay, UK
Printed in China through World Print Ltd.

16 17 18 19 20   10 9 8 7 6 5 4 3 2 1

Front cover: Snipers from 3rd Battalion, 21st Infantry Regiment, 1st Stryker
Brigade Combat Team, 25th Infantry Division provide overwatch for a fire team
at the Yukon Training Area in 2010 (U.S. Army)

Osprey Publishing supports the Woodland Trust, the UK's leading woodland
conservation charity. Between 2014 and 2018 our donations will be spent on
their Centenary Woods project in the UK.

www.ospreypublishing.com

CONTENTS

ACKNOWLEDGMENTS

As always thank you to my wife, Jodi Fraser-Neville, for her patience and support. My thanks also to my editors at Osprey, the wonderful Kate Moore and Marcus Cowper.

The author owes much to retired Sniper Master and Company Sergeant Major Nathan Vinson of the Australian Army and retired Command Sergeant Major John McPhee of the United States Army Special Operations Command. Both have truly gone beyond the call of duty in providing answers to my seemingly endless questions. As the saying goes, they have forgotten more about sniping than I shall ever know!

Thanks also to Melvin Ewing and his website, the invaluable snipercentral.com; to Mathew Coombes for his insights into the world of the police sniper; and to the handful of snipers and operators who have chosen to remain anonymous due to their chosen profession: thank you gentlemen.

LIST OF ILLUSTRATIONS

Designated Marksman of the U.S. Army observes with his Trijicon ACOG-equipped 7.62x51mm M14, Afghanistan, 2003. (U.S. Army/SFC Milton H Robinson)

U.S. Army sniper overwatching through AN/PVS-10 Day/Night Sight. (U.S. Army/SSG Joseph P. Collins Jr.)

Dutch Army sniper team fires a warning shot, Afghanistan, 2008. (ISAF/MC1 (SW/AW) John Collins)

7.62x51mm SR-25 semiautomatic rifle in a sandbagged bunker, Afghanistan, 2008. (Australian Department of Defence/Corporal Neil Ruskin)

U.S. Army Designated Marksmen with 7.62x51mm M14 Enhanced Battle Rifles, Afghanistan, 2010. (U.S. Army/PFC Donald Watkins)

Italian Army sniper training with a Sako .338 Lapua Magnum TRG-42, Afghanistan, 2010. (ISAF/SSG Romain Beaulinette, French Army)

U.S. Army Ranger providing overwatch with his 5.56x45mm SOPMOD Block 2 M4A1 carbine, Afghanistan, 2010. (U.S. Navy/MCS 3rd Class Jeffrey M. Richardson)

Dutch sniper team wearing Ghillies in the Afghan desert. (ISAF)

Australian sniper with a camouflaged 7.62x51mm Mk11 Mod 0, Afghanistan, 2010. (Australian Department of Defence/Corporal Chris Moore)

British Army sniper with a .338 Lapua Magnum L115A3, Afghanistan, 2010. (UK Ministry of Defence/Major Paul Smyth)

U.S. Navy SEAL sniper with a .300 Winchester Magnum Mk13 Mod 5, Afghanistan, 2011. (U.S. Air Force/SSG Ryan Whitney)

Australian Commandos of the Special Operations Task Group in action in southern Afghanistan. (Australian Department of Defence)

U.S. Army Designated Marksman with a 7.62x51mm M14, Afghanistan, 2011. (U.S. Army/SSG Ryan Crane)

Australian sniper on overwatch scans for insurgents, Afghanistan, 2011. (Australian Department of Defence)

British Army Platoon Marksman with a 7.62x51mm L129A1 provides overwatch, Afghanistan, 2012.

U.S. Army sniper with a .300 Winchester Magnum M2010 providing overwatch security in the Afghan mountains, 2012. (U.S. Army/Sgt Trey Harvey)

LIST OF ILLUSTRATIONS

Danish sniper firing a 7.62x51mm HK417 through a glass window. (U.S. Army/Visual Information Specialist Dee Crawford)

British Army sniper pair stalking in Ghillie suits. (U.S. Army/Visual Information Specialist Gertrud Zach)

Australian Army sniper in a "laidback" shooting position. (Australian Department of Defence)

British and French snipers in Ghillie suits. (UK Ministry of Defence/Mark Owens)

SASR sniper rapid-firing during urban operations training. (Australian Department of Defence)

Close-up of the .338 Lapua Magnum Blaser R93 Tactical 2. (Australian Department of Defence)

U.S. Air Force sniper pair in full Ghillie suits. (U.S. Air Force/SSG SC Felde)

U.S. Army Ranger with a 7.62x51mm M110 provides precision fire. (U.S. Army/SSG Russell Klika)

Special Forces sniper of the U.S. Army engaging an explosive target with his .50BMG M107. (U.S. Army/Sgt Tony Hawkins)

Green Beret zeroing his .300 Winchester Magnum M2010. (U.S. Army/Master Sgt Alex Licea)

Marine Scout Snipers line up a shot during the Mountain Scout Sniper Course. (USMC/Lance Corporal Sarah Anderson)

U.S. Marine Scout Snipers train with the .50BMG M107A1 antimateriel rifle. (Department of Defense/SSG Chance W. Haworth)

The view through a 20-power optic. (U.S. Navy/PH3 Heather S. Gordon)

U.S. Army snipers receive instruction on aerial sniper fire from a Blackhawk helicopter. (U.S. Air Force/Justin Connaher)

The cockpit of Air France Flight 8969 at Marseilles Airport, peppered with sniper shots, 1994. (Getty)

GIGN snipers providing overwatch in the aftermath of the Charlie Hebdo murders, 2015. (Getty)

LIST OF ABBREVIATIONS

ACOG	Advanced Combat Optical Gunsight (rifle combat optic)
AFO	Advanced Force Operations (low profile reconnaissance operations)
ANA	Afghan National Army
ANP	Afghan National Police
AO	Area of Operation (a military unit's operating area)
AVI	Aerial Vehicle Interdiction (helicopter sniping mission)
AWG	Asymmetric Warfare Group (U.S. military unit)
BDC	Bullet Drop Compensator (scope setting to adjust drop due to range)
BMG	Browning Machine Gun (.50 chambering)
BRF	Brigade Reconnaissance Force (Royal Marines)
BRI	Brigades de Recherché et d'Intervention or Anti-Gangs Brigade (Paris police tactical unit)
CENTCOM	U.S. Central Command (U.S. military command responsible for the Middle East and Central Asia)
COLT	Combat Observation and Lasing Team (U.S. Army surveillance and sniper team)
COP	Combat Outpost (small temporary base)
CSASS	Compact Semi-Automatic Sniper System (U.S. sniper rifle)
CT	Counterterrorist
CWIED	Command Wire IED
DA	Deliberate Action (a plan to rescue hostages at the

	timing and location of the security forces' choosing as opposed to an IA)
DM	Designated Marksman (a specially trained infantry platoon member with advanced shooting skills and optic-equipped rifle)
DMR	Designated Marksman Rifle (typically a scoped semi-automatic sniper rifle)
DOPE	Data–On–Previous–Engagements (range card listing scope adjustments for range, wind and altitude)
EBR	Enhanced Battle Rifle (modernized M14 battle rifle)
EOD	Explosive Ordnance Disposal ("bomb disposal")
EOF	Escalation of Force (responses to failure to stop incidents at checkpoints)
ESR	Enhanced Sniper Rifle (U.S. Army M2010 sniper rifle)
FAST	Fleet Antiterrorism Security Teams (U.S. Marines)
FEBA	Forward Edge of the Battle Area (location of most forward deployed friendly units)
FFP	Final Firing Point (the position from which a sniper shoots)
FOB	Forward Operating Base (semi-permanent base)
FSG	Fire Support Group (heavy weapons overwatch)
GIGN	Groupe d'Intervention de la Gendarmerie Nationale or National Gendarmerie Intervention Group (French counterterrorist unit)
GIS	Gruppo di Intervento Speciale (Italian tactical unit)
GOE	Grupos de Operaciones Especiales (Spanish counterterrorist unit)
GPMG	General Purpose Machine Gun (Belgian-designed MAG58)
GSG9	Grenzschutzgruppe 9 (German counterterrorist unit)
HBIED	House-Borne IED (booby-trapped building)
HITRON	Helicopter Interdiction Tactical Squadron (U.S.

	Customs unit)
HMMWV	High Mobility Multipurpose Wheeled Vehicle (Humvee)
HOG	Hunter–Of–Gunmen (U.S. Marine scout sniper)
HRT	FBI Hostage Rescue Team
IA	Immediate Action (also known as an Emergency Action – a plan to rescue hostages should hostage takers begin executing them)
IED	Improvised Explosive Device ("roadside bomb")
IPB	Intelligence Preparation of the Battlefield (reconnaissance and intelligence gathering)
ISAF	International Security Assistance Force (NATO forces in Afghanistan)
ISR	Intelligence, Surveillance and Reconnaissance (battlefield intelligence)
JDAM	Joint Direct Action Munition (guided aerial bomb)
JSOC	Joint Special Operations Command (U.S. military)
JTAC	Joint Tactical Air Controller (forward air controller)
KD	Known Distance (targets where range is known)
KIM	Keep-In-Memory (games to improve memory of seen objects)
KLE	Key Leadership Engagement (a Shura)
KSK	Kommando Spezialkräfte (German Army SOF unit)
LEWT	Light Electronic Warfare Teams (electronic eavesdropping and triangulation)
LMG	Light Machine Gun
LSW	Light Support Weapon (British L86A2 support weapon)
MARSOC	Marine Special Operations Command (U.S. Marines)
MFC	Mortar Fire Controller (forward observer for mortars)
MOA	Minute-Of-Angle (system for estimating size of a target or grading the accuracy of a sniper rifle)

MOUT	Military Operations in Urban Terrain
MRAP	Mine Resistant Ambush Protected (armoured vehicles)
MVD	Ministerstvo Vnutrennikh Del or Ministry of Internal Affairs (Russian internal security)
NAI	Named Area of Interest (area of potential insurgent activity)
NATO	North Atlantic Treaty Organization
NOCS	Nucleo Operativo Centrale di Sicurezza (Italian counterterrorist unit)
NOD	Night Observation Device (night vision goggles)
ODA	U.S. Army Special Forces Operational Detachment – Alpha (U.S. Army "Green Berets")
OEF	Operation *Enduring Freedom*
OIF	Operation *Iraqi Freedom*
OMON	Otryad Mobilny Osobogo Naznacheniya or Special Purpose Mobility Unit (Russian Federal Police tactical unit)
OP	Observation Post (covert surveillance position)
PB	Patrol Base (small semi-permanent base)
PBIED	Person-Borne IED ("suicide bomber")
PIG	Professionally-Instructed-Gunmen (trainee U.S. Marine scout snipers)
PIRA	Provisional Irish Republican Army (Irish terrorist organisation)
PKM	Pulemyot Kalashnikova (Russian-designed medium machine gun)
PMOT	Precision Marksman Observer Team (U.S. Customs sniper unit)
PPIED	Pressure Plate IED
PSR	Precision Sniper Rifle (U.S. SOCOM Mk21 sniper rifle)
PTSD	Post-Traumatic Stress Disorder (mental health illness)

LIST OF ABBREVIATIONS

QRF	Quick Reaction Force ("back-up" unit)
Quala	Afghan adobe building
RAID	Research Assistance Intervention Dissuasion (French police counterterrorist unit)
RCIED	Remote Control IED
ROE	Rules Of Engagement (rules governing the use of lethal force)
RPG	Rocket Propelled Grenade (rocket launcher)
RPK	Ruchnoy Pulemyot Kalashnikova (Russian-designed light machine gun)
RAR	Royal Australian Regiment (Australian infantry)
SAPI	Small Arms Protective Insert (hard plate in body armour)
SAS	Special Air Service (UK)
SASR	Special Air Service Regiment (Australian) or Special Applications Scoped Rifle (Barrett M107)
SASS	Semi-Automatic Sniper System (U.S. sniper rifle)
SAW	Squad Automatic Weapon (U.S. term for LMG, typically the M249)
SBS	Special Boat Service (UK)
SCAR	Special [Operations Forces] Combat Assault Rifle
SCO19	Specialist Command 19 (London Metropolitan Police firearms unit)
SDM	Squad Designated Marksman (another term for DM or Designated Marksman)
SDM-R	Squad Designated Marksman Rifle (another term for DMR)
SEAL	Sea Air and Land (U.S. Navy special operations)
SFARTAETC	Special Forces Advanced Reconnaissance, Target Analysis and Exploitation Techniques Course (U.S. Army reconnaissance course)
SFSC	Special Forces Sniper Course (U.S. Army)

Shura	a meeting between village elders and representatives of ISAF or the Afghan government
SOCOM	Special Operations Command (U.S. Army)
SOF	Special Operations Forces
SOTIC	Special Operations Target Interdiction Course (old name for U.S. Army SFSC)
SPR	Special Purpose Rifle (U.S. Mk12 sniper rifle)
SSE	Sensitive Site Exploitation (searching for intelligence at a target location)
SUSAT	Sight Unit Small Arms, Trilux (older British Army rifle optic)
SVD	Snayperskaya Vintovka sistem'y Dragunova (Russian-designed Dragunov sniper rifle)
SWAT	Special Weapons and Tactics (police tactical unit)
SWS	Sniper Weapon System (U.S. Army M24 sniper rifle)
TCP	Traffic Control Point
TIC	Troops in Contact (radio message denoting contact with the enemy)
UAV	Unmanned Aerial Vehicle ("Drone")
UKD	Unknown Distance (targets where range is unknown and must be estimated)
UKSF	United Kingdom Special Forces (British SOF)
UN	United Nations
VBIED	Vehicle-Borne IED ("car bomb")
VCP	Vehicle Checkpoint
VOIED	Victim Operated IED
VSS	Vintovka Snayperskaya Spetsialnaya (Russian-designed Vintorez suppressed rifle)
WMIK	Weapons Mount Installation Kit (an armed Land Rover)

CHAPTER 1
THE MODERN HISTORY OF SNIPING

The history of sniping could easily take up multiple volumes and is not the intent of this book. Instead we will simply set the scene, covering the evolution of sniping with a particular focus on the latter part of the 20th century and on events and tactics that would influence or affect the work of the modern sniper on the battlefields of the 21st century. We will briefly look at the emergence of sniping in the last century and provide a few highlighted incidents and individuals that continue to resonate through the work of today's military or police sniper.

The term "sniper" dates back to the British Empire's occupation of India in the 18th century. British officers and those of the diplomatic corps would, as a pastime, hunt a type of marshland or wading bird known as a snipe. Apparently, the snipe made challenging hunting as it was quick to spook, requiring hunters to patiently stalk their prey. If the snipe took to the air, it made a particularly difficult target as it changed course and direction on a whim. Those who became skilled hunters of snipe were thus referred to as "snipers." Prior to the adoption of this term, what we would today call snipers were more commonly referred to as "sharpshooters."

Sharpshooters had existed for nearly as long as the rifle itself. In the American War of Independence, a unit called Morgan's Sharpshooters was formed, a member of which, Timothy Murphy, performed perhaps the first sniper kill when he shot and killed a British general at a range of some 275 meters (301 yards) – remarkable for the time. The British followed suit and created an elite rifle armed unit in 1803 known as the 95th Rifles.

The significance of rifle-armed troops was that most soldiers of the day still carried smoothbore muskets that were only accurate to perhaps a third of the 300-meter average range of its rifled successor. Rifles were expensive to make and with so many muskets already produced it took until the American Civil War before rifled long guns became standard issue (rifling refers to grooves in the barrel that spin and stabilize a bullet in flight, making it both more accurate and longer ranged).

The 95th Rifles wore green rather than red (although this had more to do with tradition than camouflage), operated in loose order, making use of cover, and actively targeted leaders and runners to cripple an enemy force's command and control. The British of course suffered their own famous casualty at the hands of an enemy sniper: at the Battle of Trafalgar in 1805, Admiral Horatio Nelson was felled by a French sharpshooter positioned high in the rigging.

Hiram Berdan's Sharpshooters, formed during the American Civil War, were perhaps the first to establish a formal selection course before a soldier was admitted. Intriguingly, this apparently included the requirement to place ten rounds into a 10-inch grouping from 200 yards (183 meters). The sharpshooters on both sides during the Civil War contributed to the practice of officers retiring their swords and carrying instead the far more inconspicuous revolver. The American Civil War also saw one of the first recorded uses of what we would today term a telescopic sight: indeed a Whitworth rifle thus equipped killed Union General John Sedgwick at a range of approximately 900 meters (984 yards), a phenomenal shot with such a rudimentary weapon and optic.

The first military force to fully adopt tactics that resembled those of the modern sniper was the Boers against the British in southern Africa in 1880 and again in 1899. The British had experienced irregular warfare already, in 1838, at the hands of the Afghans and their Jezail long range rifles during the First Afghan War. The Boers, however, used the first camouflaged clothing and made use of the first smokeless gunpowder, making them very difficult to spot when deployed in concealing terrain.

In his 1906 book *Small Wars, Their Principles and Practice*, Colonel Charles E. Calwell described the disproportionate damage that could be inflicted by a sniper:

> In the first place there is the wear and tear caused by isolated marksmen perched on the hilltops, who fire down upon the troops in camp and on the march, whose desultory enterprises render outpost duties very onerous, who inflict appreciable losses among officers and men, and who thin the columns of transport with their bullets this is more prejudicial to the efficiency of the army than is generally supposed.

A unit of Scottish Highlanders known as the Lovat Scouts saw action in the Second Boer War, but it was not until their service in the Great War that they began using the gamekeeper's Ghillie to camouflage themselves. The Ghillie, a camouflage suit worn over the uniform, remains an essential tool in the armory of today's modern sniper. The Lovat Scouts (Sharpshooters) in fact became the first officially recognized sniper unit within the British Army in 1916.

The Great War saw a number of other refinements to the art of the sniper – the widespread use of quality optical sights, the deployment of trench periscopes and spotting scopes, and the development of the Hawkins Position, a prone shooting position that is still in common use today. It also saw the first recorded example of a sniper duel, the epic match between Australian sniper Trooper Billy Sing and Turkish sniper Abdul the Terrible during the Gallipoli campaign.

After the Great War, sniping took a back seat for many armies as technological advances in the field of armored vehicles took center stage. During the 1939 Winter War between Finland and Russia, however, sniping returned with a vengeance. A Finnish corporal called Simo Hayha but nicknamed "White Death" racked up an incredible 500 kills, with some claiming he was responsible for a further 200. The Soviets learned their lesson and sniping became a priority within the Soviet infantry.

During Operation *Barbarossa*, Soviet snipers harassed German forces through the inhospitable forests, swamps and marshes of the Eastern Front. Soviet partisan snipers maintained the pressure behind the lines. The Germans responded with a redoubling of effort, equipping each rifle platoon with a number of scoped rifles and training record numbers through their sniper school. One of the most famous of the Soviet snipers was Vasily Zaytsev, now immortalized in the book and later film, *Enemy at the Gates.*

Zaytsev is played by Jude Law in the film and is shown engaging in a famous duel with a German sniper instructor, Major Erwin König, played by Ed Harris. The duel began after German snipers were deployed to overwatch[1] a disputed water hole.

After the German snipers killed a number of Soviet soldiers, Zaytsev and a number of snipers were deployed in what today would be termed a countersniper role. One of the German snipers, Major König, was a master sniper sent from Berlin to hunt Soviet snipers, but he had moved to another area of the battlefield by the time Zaytsev arrived, who immediately began to stalk his nemesis.

The film is a gripping representation of the grim struggle in the ruins of Stalingrad with the two snipers, the best exponents of their craft, stalking each other in the destroyed landscape. Zaytsev finally spots his adversary, betrayed by a rookie error – a glint from the German sniper's optic – and kills him. It all makes for great cinema though unfortunately it likely never occurred, at least not featuring König. Apart from Zaytsev's own memoirs, there is no contemporary record of any master sniper named Major König serving with the German Army. Major John Plaster, world-recognized sniper historian, argues however that any record may well have been lost in the destruction of German cities toward the later part of the war. The author hopes he is right, as the story of Zaytsev versus König is a classic in the sniper literature.

The Red Army first employed two-man sniper teams, although they didn't necessarily operate in the sniper and spotter arrangement that modern snipers do. Indeed, the legendary Vasily Zaytsev even operated with another sniper. Fyodor Matveyevich Okhlopkov, a Soviet big-game hunter, and his sniping partner, Vasili Shalvovich Kvachantiradze, worked together for the duration of the war and were the most lethal sniper pair of the conflict, chalking up an incredible 644 kills. Every Soviet platoon was, at least theoretically, equipped with a two-man sniper team. Their German adversaries followed suit. At battalion level, the Russians organized a platoon of Scout Snipers who operated in a very similar manner to that of today's U.S. Marine Scout Snipers.

German snipers like Josef "Sepp" Allerberger also rose to prominence, with 257 confirmed kills on the Ostfront. Allerberger used the unusual tactic of deploying a camouflaged umbrella festooned with foliage as a kind of sniping screen to shield his silhouette from enemy snipers. He also famously halted a Soviet tank attack with a single shot, killing a tank commander who foolishly exposed his head outside the hatch. A female sniper (not unusual in the Red Army, who employed women in many combat positions), Lyudmila Pavlichenko, accounted for 36 German snipers along with over 300 other kills.

On the Western Front, British snipers were trained by veterans of Lovat's Scouts. Training included the first formal instruction in the skill of stalking, and British snipers saw active duty in Italy before the D-Day landings. German snipers proved problematic for the Allies both at the landings and inland, where they could deploy in hides carved out of the hedgerows and *bocage* of the French countryside. In fact, the German

Army issued a set of instructions or advice for its snipers in 1944 that still resonates today, according to Major John L. Plaster:

1. Fight fanatically.
2. Shoot calmly – fast shots lead nowhere, concentrate on the hit.
3. Your greatest opponent is the enemy sniper: outsmart him.
4. Always only fire one shot from your position: if not you will be discovered.
5. The trench tool prolongs your life.
6. Practice in distance judging.
7. Become a master in camouflage and terrain usage.
8. Practice your shooting skills constantly, behind the front and at home.
9. Never let go of your sniper rifle.
10. Survival is ten times camouflage and one time firing.

In the Pacific Theater, both sides also deployed snipers in an often barbaric struggle. Japanese snipers would position themselves in precarious hides, including strapping themselves into trees that, upon discovery, would inevitably lead to their deaths by return fire. The U.S. Marine Corps had established perhaps the template for the modern sniper school with its five-week sniper-training course that instructed students in fieldcraft along with marksmanship and range estimation. The Marines also began to deploy their snipers in three-man teams that prefaced modern techniques – the sniper, the spotter and flank security.

The U.S. Marines also saw the benefits of a heavy caliber long-range rifle and used their .55 Boys antitank rifles against Japanese bunkers and fighting positions. The Russians too also began to experiment in a limited fashion with using their 14.5mm PTRS and PTRD antitank rifles as sniper rifles against German positions rather than solely against tanks. This was perhaps the birth of th modern antimateriel rifle concept.

As is all too often the case, sniping was largely forgotten once relative peace returned. This attitude, shared by nearly all militaries, saw the U.S. caught wrongfooted during the early stages of the Korean War, as skilled Chinese snipers inflicted a heavy toll. Snipers with wartime experience were brought in to train new marksmen. A similar situation occurred a decade later when conventional forces began to deploy to Vietnam. Many consider the Vietnam War the birthplace of modern sniping.

Facing a canny insurgent foe, along with the regular North Vietnamese Army, U.S. Army and Marine snipers developed unique tactics such as those of Marine Sergeant Carlos Hathcock who used a .50BMG M2 heavy machine gun, the legendary ".50 cal," with a jury-rigged Unertl 10 power optic to make the longest-range confirmed kill of the war and of the century. He hit a Viet Cong insurgent at a world record of 2,286 meters (2,500 yards). His record would stand until 2002 in Afghanistan.

Hathcock was also later instrumental in developing the Marine Corps' Scout Sniper Basic Course, and a Marine sniper course was even established in Vietnam itself from 1965. U.S. Marine sniper Sergeant Chuck Mawhinney, with 103 confirmed kills and 216 probable kills, was the most effective sniper in U.S. Marine Corps history. In one engagement in 1969, he killed 16 advancing North Vietnamese Army soldiers, all at night with an early generation night-vision optic, and all with headshots, in an estimated 30 seconds using the 7.62x51mm M21.

The U.S. Army also fielded highly efficient snipers in Vietnam, such as Staff Sergeant Adelbert Waldron, who registered 109 confirmed kills, often taken from "Brown Water Navy" boats in the Mekong Delta. He killed one enemy sniper, firing from a moving boat, at an estimated 900 meters (984 yards).

Post-Vietnam, and largely through the efforts of terrorist groups and drug dealers, law enforcement agencies were forced to respond by developing specialist police tactical units, often known by their acronym SWAT (Special Weapons and Tactics). These SWAT teams saw the first organized use of police snipers and marksmen. We will cover the evolution of this unique class of sniper in a later chapter, but in essence they evolved from domestic terrorism in the United States in the early 1970s and as a response to the international terrorism that was on the rise in Europe, typified by the 1972 Munich massacre.

Sniping continued to feature in many of the brushfire wars that erupted throughout the 1970s and into the 1980s. In Beirut in 1983, U.S. Marine snipers were engaged in a deadly long-range duel with various militia snipers around the airport and the Marine barracks that were later destroyed by a suicide truck bomb. The Soviets too were embroiled in a vicious counterinsurgency war, this time in Afghanistan.

Support for sniping in the postwar Soviet Army was cyclic, as it was in many Western militaries. Their core sniper schools, that had been teaching the hard-won lessons of Zaytsev and other master snipers, were shut down in 1952 with various advanced marksmanship programs taking their place, most run at the regimental level. In Afghanistan, the platoon marksman with his Dragunov SVD (often with the bipod of an RPK light machine gun fitted for stability) was frequently drafted into the sniper role.

In the Falklands in 1982, British and Argentine snipers were employed in traditional roles, in what was perhaps the last conventional conflict of the century and perhaps forever. Argentine snipers were equipped with better night vision and a range of bolt-action 7.62x51m Steyr SSG69s, Mauser 98Ks and semiautomatic Beretta BM-59s, along with a limited number of .300 Winchester Magnum commercial Weatherbys. They proved tenacious in defense, inflicting significant casualties during British attacks; indeed, an Argentine sniper was responsible for delaying an entire company on Mount Longdon.

During the 1983 invasion of Grenada, U.S. Army Ranger snipers were used to great effect to eliminate Cuban mortar and machine gun teams, accounting for 18 enemy. Famously SEAL (Sea, Air and Land) Team 6 conducted a mission to rescue the governor general of the island, Sir Paul Scoon. The 22-man SEAL element became trapped in Scoon's residence, Government House, when it was encircled by enemy troops.

Supported by a BTR-60PB armored personnel carrier, enemy infantry began to advance. A SEAL sniper armed with a 7.62x51mm Heckler and Koch G3 SG/1 managed to virtually single-handedly stave off the attack, killing a reported 20 enemy soldiers.

In the 1989 invasion of Panama, a plan for SEAL snipers to use .50BMG antimateriel rifles to destroy Panamanian dictator Manuel Noriega's aircraft was discounted, and the SEALs were forced to conduct a costly ground assault against the airstrip at Paitilla, which killed four SEALs and wounded a number of others. During the subsequent invasion, a U.S. Army sniper team from the 82nd Airborne engaged and killed a Panamanian sniper hidden in a high-rise building teeming with civilians. The sniper was a Vietnam veteran with 38 sniper kills, and he managed a one-shot kill at 750 meters (820 yards) with his 7.62x51mm M21 (a sniper version of the venerable M14 battle rifle).

Marine and Army snipers again saw action in 1991 during Operation *Desert Storm*. In fact, a complement of .50BMG Barrett M82s were dispatched to Marine Scout Snipers in anticipation of extended ranges and the threat of Iraqi light armor. A Marine Corps sniper displayed the potency of the .50BMG antimateriel rifle when he engaged and stopped the advance of an Iraqi YW531 APC (armored personnel carrier) (or BMP-1: reports vary) at a reported 1,097 meters (1,200 yards) with two armor-piercing incendiary rounds from his Barrett.

During the 1990s, snipers from a number of NATO (North Atlantic Treaty Organization) nations were deployed to the Balkans as part of a United Nations peacekeeping force. UN troops were constantly attacked by militia snipers. French and British snipers proved particularly effective in countersniper operations. Other United Nations missions suffered similar difficulties at the same time. In Somalia, Marine Fleet Antiterrorism Security Teams (FAST) deployed sniper teams to protect the United States Embassy compound in Mogadishu. Faced with baying mobs of noncombatants shielding militia gunmen, the snipers were the only discretionary force option available.

After a number of successful U.S. shoots placing precision fire on gunmen hiding within civilian mobs, Somali tactics began to change as they learned that the Marine snipers could target them with lethal accuracy. Australian snipers deployed to Mogadishu alongside the Marines, but had less success as they were stymied by UN rules of engagement (ROEs) that governed when they could and could not shoot. One Australian sniper managed to rack up a kill, but using the issue 5.56x45mm F88 assault rifle rather than his sniper rifle when deployed as part of a dismounted foot patrol.

Somalia is probably best known from the book and film *Black Hawk Down*. As it has the most relevance to the development of modern sniping in the post 9/11 era, we will spend some time discussing the sniper-related elements of the deployment. Task Force Ranger was a U.S. Joint Special Operations Task Force that deployed to Mogadishu in 1993 in an effort to capture warlord Mohammed Farah Aideed. The Task Force was made up of the cream of U.S. special operations forces (SOF); a Ranger company; a Delta Force squadron; and a complement of Air Force Special Tactics operators along with the helicopters and aircrew of the 160th Special Operations Aviation Regiment.

A handful of SEAL Team 6 snipers also deployed with the Task Force and conducted a number of sniper missions, including several in support of the CIA (Central Intelligence Agency). They used both a relatively new sniper caliber weapon, the .300 Winchester Magnum in a McMillan stock, and examples of what would later become known as "recce rifles" – 5.56x45mm Colt carbines with longer 16-inch barrels and Leupold optics. The Delta snipers carried similarly modified carbines and customized 7.62x51mm M14s equipped with early Aimpoint red dot sights. The Delta snipers would soon use their weapons in a technique that would become commonplace within SOF in Iraq and Afghanistan a decade later – aerial sniping.

The first known use of this technique was in an aerial vehicle interdiction or AVI during one of the early missions against Aideed's finance minister, Osmond Atto. In the film *Black Hawk Down*, a Delta Force sniper effortlessly puts a couple of rounds from his M14 through the engine block of Atto's SUV. In reality, the vehicle's engine block was engaged by both minigun and sniper rifle fire from an orbiting Blackhawk. Atto managed to bail out and a foot chase ensued before he was captured by Delta operators in a nearby building. The mission however proved the concept, and AVI, and aerial sniping, began to see significant operational use.

The SEAL snipers had also been busy on an earlier mission hunting Atto. One SEAL sniper, Howard Wasdin, shot and killed a Somali RPG (rocket-propelled grenade) gunner at a distance of 846 yards (774 meters) with his .300 Winchester Magnum McMillan before he could fire on a Blackhawk overhead. The SEALs were also instrumental in positively identifying Atto after one of their snipers scaled a tower to establish a hide site overlooking a target building.

The Battle of the Black Sea (or Battle of Mogadishu) also confirmed the viability of deploying sniper teams in orbiting helicopters, while also serving as a cautionary tale of the dangers to low-flying helicopters in an urban environment. During the battle, two of the eight Blackhawks orbiting overhead carried three-man Delta sniper teams. According to a former Ranger officer interviewed by the author, the teams were deployed to provide aerial surveillance and, if required, precision fire to support the ground units.

Today much of that role has been taken on by ISR or Intelligence Surveillance and Reconnaissance assets such as armed UAVs (unmanned aerial vehicles or "drones" like the Predator or Reaper) and light aircraft equipped with high-definition video cameras to stream the action to the commander. Today assault team leaders carry a handheld monitor that allows them to see in real-time what is around the next corner or over the next hill. In 1993, that technology didn't exist and the aerial snipers played an important part in providing the Task Force commander with a view of the battlespace from above.

During the battle, two of the Blackhawk helicopters were shot down over enemy-held territory.

At one point, after the second Blackhawk was shot down and it became clear that no friendly ground forces would be able to reach the crash site before it was overrun by Somali militia, a pair of Delta snipers aboard one of the other orbiting helicopters requested permission to be inserted near the crash site. After a number of their requests were denied, as their commanding officer could not guarantee when a ground unit would arrive to relieve them, finally the two snipers were given permission, as a Ranger officer recalled:

> The last time they call, Master Sergeant Gary Gordon, the Delta team leader, got on the radio and called General Garrison and said "Sir, you've got to put us in." General Garrison said "Gary, do you know what you're asking for?" And Gary Gordon said "Yes, sir, we are their only hope."

The two snipers, Master Sergeant Gary Gordon and Sergeant Randy Shughart, leapt from the hovering helicopter (they couldn't fast rope as the rope had already been deployed earlier to insert part of the Delta assault force) and advanced on foot some 300 meters to the crash site. The helicopter was already being swarmed by militia. Gordon and Shughart fought them off and managed to extract all of the wounded crew from the downed helicopter. They then protected the crash site with expert rifle fire against waves and waves of gunmen from all sides.

Eventually they ran low on ammunition and were forced to fall back on aircrew M16s held on the Blackhawk. As even these began to run dry, the snipers used their .45 pistols. Soon after, first one and then both were fatally shot as the mob surged the crash site. Both snipers posthumously received the Medal of Honor, the American military's highest award for valor in the face of the enemy.

In 1994, the First Chechen War exposed the lack of suitable sniper training within the Russian system. Their snipers were now platoon marksmen and lacked any real fieldcraft or stalking skills. Historian Lester Grau explained, "The Russian snipers were not prepared to hunt in the ruins and to lie in ambush for days on end. The Chechens, on the other hand, knew the territory and had plenty of sniper weapons."[2] The few true snipers available within the Russian forces were from MVD and FSB (Ministerstvo Vnutrennikh Del or Ministry of Internal Affairs; Federal Security Service) special operations units.

The experiences of Chechnya forced the Russians into opening a new sniper school in 1999. During the Second Chechen War, they began using similar tactics to their adversaries, deploying small two- and three-man patrols with embedded snipers that would provide overwatch for conventional forces. This type of technique would be seen years later in Iraq as American snipers supported their infantry brethren. Other hunting parties were dispatched to specifically target Chechen snipers; these generally comprised a pair of school-trained snipers and a five-man security overwatch element, or a number of Spetsnaz snipers bolstering a detachment of marksmen.

Although the events of September 11, 2001 and more recent outrages are perhaps the first to come to mind, terrorism has afflicted Western nations since the early 1970s. Most terrorists used the tried and true methods of hijacking and bombing, with few showing the necessary skills, or bravado, to take on the security forces with sniper rifles. Indeed few terrorist organizations used sniping like the Provisional Irish Republican Army (PIRA), perhaps the best example of its use by a purely terrorist (as opposed to an insurgent) organization. Sniping was favored by PIRA as they were predominantly an urban terrorist group and could operate within a network of sympathizers that facilitated their escape after a shooting.

The PIRA would often snipe at the British Army and Royal Ulster Constabulary, inflicting a steady toll of casualties. As we shall see when we later discuss insurgent sniping in Iraq, much of the PIRA's sniping was at relatively close range, certainly under 300 meters (328 yards) for the vast majority of shots. They had access to a number of scoped bolt-action rifles, initially including .303 Lee Enfields and later the odd 7.62x54mm Dragunov SVD from Libyan arms shipments. Their most impressive weapon was the .50BMG Barrett in both its semiautomatic M82 and bolt-action M90 configurations. Incredibly, the .50BMG rifles were apparently sent through the mail in parts by supporters in the United States. The Barretts were initially employed to fire at patrols from across the border in Eire.

PIRA terrorists used urban sniping techniques like loopholing (the practice of making a small hole in a building wall to shoot through without exposing one's body), and their best exponents tended to move on after firing a single shot. Author Adrian Gilbert recounts a troubling report of a PIRA marksman placing a round through the rear window of a Saracen armored personnel carrier, killing one British soldier and wounding two. Remarkably, the terrorist had fired from within a house, aiming out through the letterbox slot in the door.

British Army snipers responded to the terrorists, becoming experts in building and maintaining covert OPs (observation points) or hide sites, often manning their hides for weeks at a time. Their hard-won experience would assist in keeping snipers alive many years later in both Iraq and Afghanistan: indeed many of the drills of urban sniper overwatch had their genesis in Northern Ireland. The snipers would often deploy disguised among a regular Army patrol who were conducting building searches. When the patrol left, they were two to four men short, as the snipers would take up residence in the attic, unbeknown to the occupants. Much of the snipers' work in Northern Ireland was surveillance rather than kinetic operations, as shooting people is currently termed.

There were "kinetic operations" too, however. A large proportion of the terrorists shot by non–Special Forces were by snipers from the British Army and Royal Marines. In one incident after a PIRA ambush of a Ferret scout car in 1972, a Royal Marine sniper pair fired 83 rounds between them, hitting ten terrorists, including one shot at an incredible

1,344 meters. With primitive optics, only a 4 power magnification and no laser rangefinder or modern aids, that was incredible shooting. Other operations were mounted by UK Special Forces to target PIRA marksmen.

In one famous operation, the British SAS captured a four-man PIRA Active Service Unit that had been responsible for the murder of nine members of the security forces. The terrorists had been using one of the .50BMG Barretts and had developed an ingenious method of covertly firing it. The terrorist marksman was placed in the trunk of a Mazda sedan with a loophole to fire from, and even makeshift armor should he receive return fire. Similar techniques were copied by criminals such as the Washington Beltway snipers in 2002 and, as we shall later see, by Iraqi insurgents who developed their own variations on the PIRA technique.

CHAPTER 2
THE MODERN SNIPER

ROLE

Today's sniper is far from the grim-faced loner of innumerable World War II films who haunts the enemy front line, notching up kills on his rifle stock. Instead, many would argue that the sniper's real role, and his most valuable one, is as a reconnaissance asset. The modern sniper team, and it is almost universally now a two- or three-man team, has the skills and techniques necessary to covertly insert into an area of interest, construct an observation post and stay unseen for days or weeks at a time, all in environments as varied, and as inhospitable, as the mountains of Afghanistan or the urban jungles of Baghdad.

One British sniper noted, "In Basra, we were the British Army's eyes on the ground. We would take up a position in a tall building and log the number of enemy soldiers and weapons we could see."[1] A U.S. Marine Corps Scout Sniper instructor agreed: "A UAV is going to be able to report vehicles or whatever the case may be but that Marine on the ground observing through those optics is going to be able to make out somebody who seems nervous or seems out of place."[2]

Other veteran snipers view their kinetic role as still their primary one. "Their first mission is to provide long-range precision fire, anywhere from 600 yards to over 1,000 yards. Our second mission is to observe and report what we see on the battlefield to the commander. If the target presents itself, we then engage and get rid of it,"[3] argued a sniper instructor with the U.S. Army's 36th Infantry Division with a handful of Iraq deployments under his belt.

When snipers do shoot, they have a disproportionate effect on the battlefield. In terms of purely tactical impact, a single sniper team can suppress a much larger force, often for extended periods of time. Most readers will have read or seen in films the typical World War II scenario of a well-placed sniper halting the advance of an infantry platoon or company. Despite advances in technology and training, this still remains the case in modern warfare, with even insurgent snipers, in reality hardly marksmen, holding up much larger and better trained forces. Accurate single shots, along with the omnipresent threat of IEDs (improvised explosive devices), can often better suppress advancing units than direct small arms fire and rocket-propelled grenades.

In those same World War II movies or books, the sniper will often be located in a church bell tower, as in *Saving Private Ryan*, or a similarly obvious hide location. Competent commanders would ensure that any potential sites that might conceal a sniper or an artillery forward observer received some speculative high-explosive fire at the very least. Even with today's far more restrictive rules of engagement (the conditions under which coalition forces can open fire on the enemy), units will generally be ready to suppress any such likely firing points the moment an insurgent sniper fires.

Along with their undoubted tactical impact, snipers also have a deep psychological effect on the enemy. The fear engendered by the knowledge that a sniper is operating in your vicinity can be debilitating, even for trained and experienced soldiers. Imagine the psychological impact of coalition forces snipers on untrained and inexperienced Iraqi insurgents – the coalition snipers were rightly feared, sometimes to semi-mythological proportions. That fear translated into a distinct restriction upon what the military terms the enemy's "freedom of movement." Put simply, the bad guys were too scared of the snipers to feel confident breaking cover and moving around, even at night.

For these reasons, the modern sniper is commonly referred to as a "force multiplier." A force multiplier is something which, when added to a military unit, multiplies its effectiveness on the field of battle, usually out of all proportion to its relative size. A sniper pair increases both the lethality and engagement distance of an infantry platoon, as it provides a combat capability it otherwise would not have. As we will see later in this chapter, the force multiplication effect provided by snipers in Iraq and Afghanistan led directly to the implementation of a program to develop what would become known as Designated Marksmen.

The idea was that with the addition of a Designated Marksman (DM), known as the Platoon Marksman within the British Army, some of what the sniper brings to the table could be provided by sharpshooters within the infantry platoon. In fact, the idea resurrected a German World War II tactic of training several men within the platoon as marksmen and equipping them with scoped rifles. Of course the Russians had long seen the relevancy of platoon- and section-level marksmen, as we have noted earlier.

The U.S. Army defines the standard nine-man infantry squad as "capable of establishing a base of fire, providing security for another element, or conducting fire and movement with one team providing a base of fire, while the other team moves to the next position of advantage or onto an objective." It can only, of course, conduct these tasks while suppressing or targeting the enemy out to the effective range of its longest range weapon.

Until the advent of the Designated Marksman, this was typically around the 300- to 400-meter mark. The only commonly carried weapons platform that could engage at longer distance was the 7.62x51mm

M240B medium machine gun. The concept of the Designated or Squad Designated Marksman aimed to change that and give infantry squads back a weapons platform that could engage point targets with accuracy to 600 meters or more. We will cover the development of the concept in some detail in a later chapter.

In support of conventional operations the U.S. Army list the following as missions suitable for snipers: "movement to contact, attack of built-up or fortified area, river crossings, support of reconnaissance and combat patrols, extended ambushes, cordon operations, deception operations." For most of these mission types, the sniper will be providing sniper overwatch, either attached directly to the infantry unit and covering its movement, or as part of a fire support position that is located on high ground overlooking the battlespace.

An American combat infantryman writing for the excellent thefirearmblog.com explained:

> Sniper teams are typically used either in support of an attack, by locking down lanes of potential retreat, or as observation posts, to watch for enemy movement either in or out of the objective. They can be used in conjunction with mortar teams, the snipers calling for fire as they observe enemy movement – this allows them to kill the enemy without revealing their positions by firing.[4]

We will examine the specific roles of snipers in both Afghanistan and Iraq in the following chapters, but before we delve much further, it is worth taking a moment to identify a few areas that may cause confusion as we discuss modern sniping.

Distances

One of the key difficulties in writing about small arms and snipers in particular is that for many years the U.S. military tended to measure ranges, particularly in terms of sniper platforms, in yards, while the rest of the world has tended to do so in meters. It leads to a somewhat jarring mixture of measurements. In the interests of authenticity, I have used both but have added the approximate equivalent distance in parentheses following the quoted range. Ironically, the U.S. Army has recently adopted metrics with the procurement of its latest sniper rifle, the M2010.

Calibers

Another confusing aspect comes from weapon calibers and how the diameter of the bore is measured. Most military calibers, but certainly not all, are measured in millimeters, while most civilian calibers, or calibers originally of civilian origin, are measured in inches. A 7.62x51mm round is thus a .308 in the civilian world. The second number after the bore diameter is the length of the casing: thus one can see that a 7.62x51mm has a longer case, and consequently probably more powder, than a shorter 7.62x39mm. In this book we will use the most commonly referenced military term for a particular caliber.

Mildot Versus MOA

Optics is the third confusing element when people start to talk about sniping. Scopes are manufactured in two main types: those that display the reticle (the cross hairs inside the optic or, as the NATO sniper school prefers, "a film with printed lines inside a telescopic sight used to aim a weapon") in mils or mildots, and those measured in MOA or Minute of Angle.

The mildot was originally an innovation of the U.S. Marine Corps, designed to provide a simple system of range estimation for its snipers in the days before laser rangefinders. It uses the Milradian as its unit of measure. One Milradian is, for the sake of our discussion, equivalent to 3.6 inches at 100 yards (91 meters).

MOA equates to just over 1 inch at 100 yards (91 meters). Both units of measure allow the sniper to estimate the size of a target at a particular range and are used for altering the optic's elevation or windage to account for range and wind conditions. Both systems have their adherents, although mildots is now far more common in military use, while police shooters often still use MOA.

Briefly, there are two methods to compensate for crosswinds and bullet drop. One is to use the DOPE (Data-On-Previous-Engagements) or range card to make adjustments on the scope using the elevation and windage knobs (vertical and lateral adjustments). The other is to use either the mildots or MOA markings within the reticle to judge the amount of hold-off needed (or how far above or to the left or right you will need to aim to hit the target). We will cover this in more detail later in the chapter.

THE FUNDAMENTALS

Before we begin to discuss the ways different militaries today select and train their snipers (and designated marksmen) and how snipers are organized, we must discuss and explain a few technical terms, and a few not-so-technical terms, that will appear throughout the book. We'll also take a look at the anatomy of a shot and the effects of a successful hit upon a human target.

What is a Sniper Rifle?

A sniper rifle is simply any rifle that is adopted for use by a sniper. Typically it will feature a magnified optic known as a telescopic sight or scope. This optic may be either fixed or variable strength magnification. A fixed magnification optic has only one setting, so for instance a 10 power scope will magnify objects ten times when viewed through the scope. Far more common with modern sniper rifles are variable magnification optics. The magnification on such scopes can be chosen by the sniper. They offer far more flexibility than a fixed magnification optic.

Another common feature of a sniper rifle is a long barrel, typically between 24 and 25 inches (609mm to 635mm), as this both increases the muzzle velocity of the bullet (the speed at which the bullet leaves the barrel) and helps stabilize it before it leaves the muzzle. Compact sniper rifles with far shorter barrels do exist, as we shall discuss in a later chapter, but these are designed for specific purposes such as covert operations or urban combat. As velocity and stabilization increase, so does the effective range, and in some respects the lethality, of the bullet. This is why sniper rifles are generally longer than the average assault rifle carried by the sniper's infantry colleagues.

The barrel itself is often specially designed and produced. These are known as match grade barrels. Most modern sniper rifles also feature free-floating barrels, as they provide a consistent point of impact across a range of shooting positions. Free-floating simply means that the barrel is only attached to the actual receiver of the weapon. A free-floating barrel will maintain the rifle's point of impact as long as the barrel is not knocked. It also helps reduce the impact on accuracy from the barrel heating up from repeated firing.

Most dedicated sniper rifles, as opposed to modified commercial weapons, will have any number of additional features that make them suited for military use. Integral folding bipods to provide a stable shooting platform are common, as are side-folding stocks that make the weapon far easier to carry. The majority of current sniper rifles will also be equipped with a sound suppressor, often erroneously referred to by the media as a silencer. Far from fully eliminating the report of a sniper rifle, the suppressor assists in two principal ways – reducing both the noise of firing and the muzzle blast to confuse the enemy as to the true location of the sniper. It also has the added benefit of preserving the hearing of the sniper team and any fellow soldiers nearby (although most now wear some form of hearing protection in combat).

The accuracy of the sniper rifle is generally rated in terms of MOA or Minute of Angle (even rifles that use mildot scopes). This is a way of measuring the fall of shot from a particular rifle and ammunition combination. The lower the MOA, the closer the grouping and thus the more accurate the rifle is considered. 1 MOA translates as 1 inch at 100 yards (91 meters). If a sniper rifle is consistently able to put rounds within a 1-inch circle at 100 yards, then that particular rifle and ammunition combination is termed as 1 MOA capable.

The MOA also increases by an inch every 100 yards (91 meters) as the range to the target itself increases. For example, at 300 yards (274 meters), the rifle would need to be placing its rounds consistently within a 3-inch circle for the rifle to still be shooting 1 MOA. At 1,000 yards (914 meters), the weapon would have to be able to place its groupings into a circle no larger than 10 inches to be considered 1 MOA. Most general-issue military sniper rifles are capable of 1 MOA or sub 1 MOA shooting. Some specialist counterterrorist, SOF (special operations forces) and police rifles are rated as .5 MOA; the grouping of such a rifle needs to be falling within half an inch at 100 yards (91 meters).

The optic mounted on the sniper rifle is nearly as important as the rifle itself. Along with providing a means of identifying a target, the scope can counteract the two principal factors affecting a sniper's bullet that he or she must attempt to counter, namely range and wind. Most telescopic scopes have at least two turrets that allow for adjustments to be made. One of these adjusts for elevation and thus is used to try to compensate for bullet drop across the distance to the target (bullets do not travel in straight lines and the trajectory increases as the distance does). The other is for windage and, as the name suggests, is used to counter the effects of crosswinds on the bullet and make lateral changes. We will explain both later in this chapter.

What is Fieldcraft and the Stalk?

Fieldcraft is the set of unique skills that a sniper and his team will use to ensure they are concealed from easy discovery while they are either watching or approaching their target. It covers the art of camouflage and concealment; the techniques of building an observation post, often termed a *hide* or *hide site*; even how to move without drawing attention to oneself by using natural dips in the earth to conceal one's approach.

The stalk, taken from the civilian hunting term meaning "to pursue prey," is the term snipers use to describe that process of stealthily approaching the target.

Modern snipers may insert, or infiltrate to use the correct terminology, by any manner of means depending on their role and the target of their infiltration.[5] Often these are extremely dangerous infiltrations, but they are usually far removed from the traditional image of the sniper in a Ghillie suit crawling slowly toward his distant target.

There is still a need, however, for the traditional skills, including use of the Ghillie. The Ghillie suit is a sniper tradition, handmade by each individual to suit his unique requirements and the environment he will be operating within. As we have seen earlier, the Ghillie originated with the Lovat Scouts in the Second Boer War and the Great War, the first true sniper unit and the first to use the Ghillie of Scottish gamekeepers on the field of battle. The design and construction of the Ghillie has changed little since.

"There is no standard Ghillie. Each is hand built to reflect the personality of its owner and the environment it will be used in," explained former FBI Hostage Rescue Team (HRT) sniper Christopher Whitcomb in his insightful book on HRT sniping, *Cold Zero.* "I spent days weaving mine together from burlap strips, cargo netting, unbraided hemp, and tree-bark colored Cordura. When it was done, I dragged it behind a truck through the mud [and] rubbed deer shit all over it." Military snipers in Iraq and Afghanistan have continued to use the Ghillie where appropriate, particularly when deploying into rural hides.

An instructor with the SFSC (Special Forces Sniper Course) added that he used a Ghillie in Iraq:

I used Ghillie suits some in Samarra, Iraq but it wasn't for the traditional stuff like hiding in brush and camouflaging myself. I had a Ghillie net I would drape over my hat [helmet] with my NODs [Night Observation Devices or night vision goggles] down, and it broke up my outline so it didn't look like a helmet. I also had a really nice Ghillie suit that I had made ... that broke up the outline of my shoulders ... I kind of blended in as opposed to giving that nice sharp outline.[6]

Fieldcraft extends to the urban environment where many of today's wars are fought. Sniping in such an environment inevitably brings to mind images of insurgent snipers fighting in the ruined streets of Beirut or Aleppo. These men are, at best, barely marksmen. Real snipers, those trained in the best of the sniper schools, would never be seen, the only evidence of their presence the effects of their deadly shooting.

They use techniques like loopholing to punch unobtrusive holes in walls to allow them to target the enemy (indeed, experienced snipers will carry a crowbar or similar implement to dig out a loophole). They will never expose themselves at an open window or doorway but instead use the darkness at the back of a room for concealment. They will also sometimes use a form of urban Ghillie, a coverall camouflaged in the colors of stone and brick to further blend in with their location.

Fieldcraft also encompasses building a hide site. The hide site is the camouflaged shooting position or observation post that the sniper will operate from. There are three principal types of hide: the expedient position, the belly hide and the semi-permanent. The expedient is simply that, a quickly constructed shell scrape in existing foliage or other concealment where the sniper is relying on his personal camouflage to conceal him rather than a purpose-built hide. The belly hide adds a roof to the sniper's position, giving him all-important shadow to operate from. Finally, the semi-permanent hide, which almost approaches a makeshift bunker, is a position constructed to allow the sniper team to move around inside hidden from view and typically has a dedicated sleeping area. There is a fourth kind, the permanent hide, although this typically takes the form of a defensive position or sangar that is located on the outer wall or main gate of a coalition forward operating (FOB) or patrol base (PB).

In operations in urban environments, there are a number of typical hide locations that the sniper can use: the room hide, the attic hide, the crawl space hide and the roof hide. Each has their own advantages and disadvantages. For instance, the crawl space is perhaps the most covert and may withstand searches by enemy troops or insurgents of the house in which the sniper is concealed, making it a favorite for SOF. Former British sniper instructor Mark Spicer[7] uses the mnemonic of COCOA to assist in selecting a hide or an OP site:

C- Cover from view and fire
O- Observation of approach routes
C- Covered approach and exit routes
O- Observation arcs as wide as possible
A- Alternative positions

SEND IT! HOW SNIPERS ENGAGE TARGETS

To understand how a modern sniper engages a target, let's look at a basic scenario, not dissimilar to those practiced at sniper schools the world over. A sniper and his partner, an observer known as a spotter, must infiltrate into what is known in the trade as a Named Area of Interest (NAI) to locate an individual who has been designated as an enemy. If the opportunity presents itself, the sniper is to take the shot and attempt to eliminate this individual.

The role of the spotter or observer in any such scenario is as important as that of the actual sniper. He will often locate the initial target, estimate the range to that target through either a laser rangefinder or via a manual process that we will cover later, and issue any corrections to the sniper after watching the fall of the shot (literally where the fired round hits). He can also act as a secondary shooter, providing another man on the trigger should the team be faced with multiple adversaries. Finally, he is also responsible for the physical security of the sniper team while the sniper is "on the gun," although incidents in Iraq and Afghanistan have now changed the way security is managed in the field (indeed, sniper teams are now often accompanied by a fire team or more infantry as force protection).

Often, the spotter is the more experienced and veteran member of the team. The U.S. Army's Special Operations Target Interdiction Course (now renamed the Special Forces Sniper Course) manual notes:

> When employed, the more experienced in the pair will act as the observer during the shot. This is especially important on a high priority target. The more experienced sniper is better able to read winds and give the shooter a compensated aim point which takes into account the effects of the environment, and consequently, better ensures a first-round hit.

The manual expands further on the topic:

> The coach–shooter relationship of the sniper pair is invaluable in target acquisition, estimation of range to targets, observation of bullet trace and impact, and offering corrections to targets engaged. Additionally, the mutual support of two snipers working together is a significant morale factor during employment in combat environments or extended missions.

The sniper must deploy in the most comfortable and most practical shooting position in preparation for taking the shot. Surprisingly, there are a large number of recognized sniper shooting positions developed for shooting in all kinds of environments, but we will only cover a handful of the most common here. Readers who wish to enhance their understanding of this and many other elements of the sniper's craft from a visual perspective are directed to Sergeant Major Mark Spicer's *Illustrated Manual of Sniper Skills* for a solid picture of sniper techniques and tactics, and/or Major John L. Plaster's illustrated and superlative *The Ultimate Sniper: An Advanced Training Manual for Military and Police Snipers*, which illuminates pretty much anything an enthusiast might want to know. Book publishers have flooded the market with alleged "how-to" books on the topic, but these two illustrated books were written by acknowledged veterans and experts in the field and are highly recommended as they show, through a series of detailed diagrams, examples of specific shooting positions.

In terms of firing positions, the *prone supported* is the most common. This is the most stable of firing positions, supporting the weapon with both hands and most typically with a bipod. The Hawkins Position is similar but is designed to conceal the sniper as much as possible and is used extensively in rural desert hides as it exposes the lowest silhouette to any watching enemy. It requires the sniper to occupy a natural dip in the ground or to construct a *shell scrape* to allow his rifle to be steadied by the ground itself.

One of the oldest but still commonly used shooting positions is the *prone backward* or *laidback* shooting position. This looks awkward but is in reality both conducive to accuracy and rather comfortable. The sniper lies on his side and brings his knees together, supporting the rifle across his thighs. Tactically it presents far less of a target than kneeling or standing and is obviously useful if the prone position cannot be adopted. The final common firing position is the *sitting* position. This may be accomplished with the legs crossed or legs apart with the sniper's elbows resting on his knees. Each change to a sniper's firing position will have an impact upon his point of impact downrange.

Once the sniper is in a comfortable shooting position and has a view of his target he must estimate or measure the range. This is often the task of the spotter, but the sniper must also be skilled in accurate range estimation. If he has been in place observing the target area for some time, he will spend that time developing a range card that shows his range estimations to major landmarks in the area to allow him to either know or accurately guess the range (often referred to as the *yardage*) to any target located near one of these previously ranged points. He will also make a sketch of the area, both to aid in range estimation and for intelligence-gathering purposes. Such a sketch will also be invaluable when conducting an after-action review following the sniper operation or, as Nathan Vinson, former Master Sniper at the Australian Army's School of Infantry, adds: "if a relief-in-place of another sniper team for the continued observation of the target area is required."

Range estimation can be done manually by using a number of methods. One is simply based on the sniper projecting a known range (for instance 100 meters) and visually extending that known range in increments to the target. Another is the known distance method where the sniper estimates the range based on the distance to an object that he already knows (this is where a range card showing the range to key landmarks can be invaluable). Another is to use the mildots in his optic (if he is so equipped) and a relatively straightforward equation to estimate the distance to a target.

Vinson takes us through how this works in detail:

There are two main methods of judging distance without aids. They are: a) by the unit of measure; and b) the appearance method. To use the unit of measure method, [the sniper must] visualize a known distance on the ground and calculate how many of the units would fit between the observer and the object. An easy figure to use is a unit of 100 meters. This method gives acceptable results when: the observer can see all the intervening ground; and the distance to be estimated is not greater than 400 meters.

The appearance method is based on what an object looks like compared to its surroundings. To become proficient in judging distances by this method a great deal of practice is required, under varying conditions of ground and observation. The amount of visible detail of a person at various ranges gives a good indication of the distance he is away. An observer with good vision should be able to distinguish the following detail:

a)   At 100 meters: clear in all detail;
b)   At 200 meters: clear in all detail, color of skin and
        equipment identifiable;

c)   At 300 meters: clear body outline, face color good, remaining
     detail blurred;

d) At 400 meters: body outline clear, remaining detail blurred;

e) At 500 meters: body begins to taper, head becomes indistinct; and

f) At 600 meters: body now wedge shape, no head apparent.

There are a number of aids that can assist in judging distance. These include:

Bracketing: the method most likely to prove the best under all conditions. The observer should decide on the furthermost possible distance and the nearest possible distance to the object. The average of these is taken as the range. For example, if the furthest estimated distance is 1,000 meters and the nearest distance is 600 meters then the range is therefore 800 meters.

Halving: useful for judging distance up to 1,000 meters. The observer estimates the distance to a point half-way, and in a direct line to the object, and then doubles it. The main disadvantage of this method is that any error made in judging the distance to the halfway point is obviously doubled for the full distance.

Key Ranges: when the range to any point within the [sniper's] arc of observation is known, the distance to another object can be estimated from it. This method can be successful provided that the object is reasonably near the key range object.

Unit Average: provided that there is sufficient time available, the observer should get several other soldiers from the team to estimate the distance to the object. He should then take the average of their answers. If all the sniper team are practiced in the skills of judging distance this method can be particularly accurate.

Binoculars: can be used to estimate distance, particularly at long range by using the sub-tension rule provided the height of an object is known. If an object is known to be 4 meters high and it is exactly covered by the smallest graticule (the measuring scale within the optic), it will be about 1,000 meters away. If the object is 8 meters high, then it will be 1,000 meters away if it is exactly covered by the small graticule. This is similar to the method outlined earlier, using the mildots in the scope to estimate distance.

Laser Rangefinder: the rangefinders can have a maximum theoretical range of 10 kilometers, however, atmospheric conditions will often limit the maximum range to 6 to 7 kilometers. A ranging accuracy of 5mm is possible however regardless of the distance measured.

Although all modern snipers use laser rangefinders, all snipers learn the manual methods in case the device is broken or runs out of batteries, in a similar manner to learning how to use a compass even though they will always be equipped with a GPS receiver. Indeed the sniper or spotter can use a map to approximate the distance between two points using the distance scale of the map. Once the range is known, the sniper needs to begin making adjustments based on that range and on environmental factors.

As we have noted earlier, bullets do not travel in a straight line. They actually travel in an arc to their target and the longer the range, the greater the arc. The sniper needs to understand what the trajectory of his bullet looks like at various ranges so that he can compensate for the drop of the bullet due to gravity and air resistance as it arcs through space (when snipers talk about a *flat shooting* round they are talking about a bullet whose velocity is high enough at a given range to counter this bullet drop to some extent).

Nathan Vinson adds: "The sniper also needs to be aware of the culmination point of the round. If there is an object between the sniper's direct line of sight to the target and is at the point of where the round reaches its highest point, or culmination point, then the round may hit that object and not the target."

The sniper has to make an adjustment for drop by using the elevation turret on his scope to dial in the range, taking into account the projected drop of the round. Snipers will typically have a range card (also known as a DOPE card for Data-On-Previous-Engagements) that will list the drop at various ranges using different bullets, although some nations eschew the use of the cards and believe the sniper should be trained to make accurate estimations each time he fires.

Most modern scopes feature what is known as a bullet drop compensator (BDC) that is designed for an express caliber and will show the necessary adjustments required in the reticle of the scope. An optic featuring a bullet drop compensator will show a number of aiming points, with the central aiming point being whatever the rifle is zeroed at. The others are typically shown in 100-yard (91-meter) graduations so the sniper can quickly balance the effect of bullet drop once he has estimated the distance to his target. Unfortunately this can never be totally accurate as BDCs cannot take into account the type of round fired, crosswinds, temperature or altitude. They do however provide an important indicator that the sniper can use to begin to adjust his shot to ensure a hit.

Bullets themselves also move away from the rifle's bore line due to an effect known as *spin drift*, which is produced by the rifling in the barrel that also stabilizes the bullet. Spin drift, like most factors affecting shooting performance, becomes more pronounced at longer ranges. A sniper will zero his rifle to know how the trajectory, or bullet drop, and the spin drift will affect his shooting at different known ranges. To do so all sniper rifles must be zeroed, or calibrated by the sniper, for a particular distance.

Once zeroed and with all other things being equal, the bullet should strike exactly where the sniper aims through his scope at the zeroed range. The zeroed range is the most likely distance the sniper expects to be firing at. In Iraq, zeroing for 400 meters (437 yards) was common, while in Afghanistan 600 (656 yards) or even 800 (874 yards) meters was used. Some units zeroed for much closer distances, such as 100 meters (109 yards) and then confirmed that zero at every 100-meter (109-yard) increment, as it made adjusting for distance faster. Snipers will also zero their weapons both with and without suppressors mounted, as they will affect the point of aim. The sniper would discuss with his team leader or the patrol commander how a specific operation would be conducted and what was likely to be required of the sniper team before the sniper settled on a range for which to zero his weapon.

The sniper (or spotter) should also measure the air temperature (as warmer air makes bullets fly both faster and flatter in terms of trajectory) and the humidity of the air, as this will affect the flight of the round as well. The denser the air, the lower the round will fall on the target due to the resistance. He will then take into account the altitude of his FFP or Final Firing Point (the location from where the sniper will take his shot), and what direction the wind is flowing, bearing in mind that depending on the topography of the target area, and the season, the wind may be moving in several directions and at a number of different speeds.

Crosswinds can easily shear a bullet from its target, and like most factors affecting sniper shooting, the longer the range, the more the impact. Many readers will also remember *Coriolis drift* or the *Coriolis effect* from console games such as *Call of Duty: Modern Warfare*, where a sniper must account for the gravitational effect of the rotation of the Earth on his bullet. In reality, this only comes into play at really extreme ranges.

Thankfully, many of these considerations can be factored in for the sniper by the use of increasingly effective ballistic computers. These are now even available for iPhones and iPad tablets, and have become standard issue for most Western snipers. A typical app will calculate trajectory at a given range, the effects of wind (windage), temperature, attitude and barometric pressure factors along with air density.

More advanced versions contain a database of specific projectiles so that the sniper or spotter can specify the load they are shooting and the app will take it into account in its calculations. Vinson notes that with all such technological solutions, "you will [still] have to add this data to the calculation, however this only calculates factors from the shooter's end, not at the target end. It increases your chances of getting a first round hit but doesn't guarantee it."

Finally, the sniper will have to consider any movement on the part of his target and compensate for this. There are two general methods to do this: tracking, which follows the target in the reticle of the optic before leading the shot (adjusting his point of aim using mildots in his reticle to give the bullet time to arrive so normally aiming ahead of the moving target), or the ambush or trap method. To accomplish this, the sniper will also have to be able to judge the rate of movement: is the target walking or running and at what approximate speed?

This latter method sees the sniper place the reticle of his scope ahead of a known target and allows the target instead to move into the sniper's field of vision, firing when the sniper judges there is enough lead to hit or using the mildot method to judge the speed of his target and accounting for further environmental conditions like wind.

All of these factors will influence the sniper's final hold-off or where he places the reticle of his optic in relation to his target. The sniper will ensure that his weapon is firmly grounded using either an integral bipod/tripod and/or a sand sock or sandbag carried just for this purpose to eliminate any physical movement that will translate to the rifle. The weapon's sling may also be used, firmly secured to either the arm or leg of the sniper to aid in supporting the rifle and shooter.

The sniper must also endeavor to have the optimal cheek weld between the stock of his rifle and his cheek. He does this to ensure two things: one, that he has enough eye relief from the scope to stop it from recoiling back into him and to ensure that his field of view remains unchanged, and two, so that his head and the rifle recoil together, allowing for a rapid recovery and follow-up shot as required. Snipers try to use the same cheek weld every time they shoot, as a change in cheek weld can negatively affect the point of impact of his rounds.

Snipers never pull or snatch at the trigger; they squeeze the trigger. Any movement can cause vibrations into the action of the weapon that can have a negative effect on accuracy. Snipers practice breathing to steady their nerves and reduce any detrimental impact on their shot.

Vinson adds that "the rule of thumb is to release the shot when there is no air in the lungs."

Most dedicated sniper rifles have a consistent trigger pull of around 2.5 pounds (just over a kilogram), although many modern designs allow the sniper to adjust the trigger pull or pressure to suit their individual preferences. The sniper will take up the slack on the trigger with the tip of his trigger finger and breathe out, expelling all the air in his lungs and relaxing his muscles. At the end of the breath, he evenly squeezes the trigger, firing a round that should hopefully hit its target.

The spotter, using a spotting scope or similar magnified optic, will *call the shot*. This means he will provide direction to the sniper on where the shot struck and, if it missed, adjustments to bring the next shot onto target. "Calling the shot is being able to tell where the round should impact on the target. Because live targets invariably move when hit, the sniper will find it almost impossible to use his telescope to locate the target after the round is fired," explained Vinson, so the spotter must use the far wider field of view of his spotting scope to call the adjustments.

EFFECTS ON TARGET

Despite what may be portrayed in Hollywood thrillers, modern snipers tend to aim for the largest target, the region between the top of the legs and the shoulders, which is the so-called center of body mass. This gives the greatest probability of a hit, and, while not as certain of inflicting an immediate kill, a high-velocity rifle bullet striking anywhere in this area will cause even the most fanatical opponent to pause. If the round strikes any of the vital organs contained within this zone, the target will expire shortly afterwards.

The center of body mass is also where any body armor is likely to be located, with the hardened ceramic or steel trauma plates covering those vital organs. Although insurgents and terrorists, the most likely targets of today's snipers, are rarely equipped with body armor, the sniper must be capable of aiming for the headshot if required. Most military snipers will generally only attempt headshots against static targets out to 200 meters (218 yards) unless the situation dictates otherwise (for instance if the head is the only visible target).

The headshot is nearly always instantly fatal. The sniper tends to aim for what is known as the *fatal T*. The fatal T is a region of an enemy target's head with the crosspiece of the T across the eyes and the vertical part of the T centered down the middle of the face. A hit within this fatal T will destroy the brain stem and will instantly kill the target. This shot is particularly important when dealing with suicide bombers or hostage takers as it will cut all brain signals to the body and stop any final hostile acts. The author has unfortunately seen the effects of a solid strike to the fatal T on a number of combat videos and can attest that the combatant simply collapses to the ground like the proverbial rag doll.

No matter where they strike, rifle bullets create two wound channels in the target. The first and most important is the permanent wound cavity. As its name suggests, this is the channel carved by the path of the bullet, and any fragments of that bullet as it breaks apart, through a human target. The wider this permanent wound cavity, the greater the chance that the target will simply collapse from massive tissue and organ damage. To create a sufficiently wide permanent wound cavity, the larger and faster the bullet the better, as, if the projectile doesn't fragment or expand, the size of the wound is closely related to the diameter of the bullet.

The temporary wound cavity is the channel produced by the disturbance of the flesh, muscles and organs by the creation of the permanent wound channel – the shock effect around the wound, if you will. It is sometimes referred to as "hydrostatic shock." This is purely temporary in nature, as the disturbed tissue, muscles and organs will move back into place after the passage of the bullet or bullet fragments. A temporary wound cavity is only produced by a high-velocity round and is thus not typically seen in wounds inflicted by pistols or submachine guns. A sufficiently major disruption to the tissue and organs will increase the severity of the wound.

Readers may also have read of entry and exit wounds. Rifle-caliber projectiles tend to produce a very small entry wound, but, if they penetrate through the body, a much larger exit wound. Even match-grade sniper rounds designed to expand are unpredictable when they strike a human being and numerous factors come into play, which may see a *through and through* wound produced (this is when a bullet has passed through the body without fragmenting or striking any major obstructions, often more likely with military-issue ball or full-metal-jacket ammunition as it is designed to penetrate not expand).

In some circumstances, such wounds may prove dangerous to hostages or noncombatants in the immediate area, as they may be hit by the projectile as it overpenetrates its original target. This can also be in the sniper's favor, such as when, famously, a British sniper in Helmand struck a Taliban insurgent on a motorbike with a round from his .338 Lapua Magnum L115A3 rifle. The bullet went through its original target, killing him, before striking the Taliban riding pillion, also killing him. An American sniper in Iraq also managed to kill two insurgents with a single round when firing his .50BMG Barrett M107.

For a reliable chance of stopping an enemy from carrying out hostile acts, the round needs to penetrate 12 to 18 inches, dependent on where on the body the individual is hit and his or her physical size and mass. For immediate incapacitation, a shot must strike and sever the central nervous system, most commonly accomplished through the headshot. Otherwise, a strike that penetrates sufficiently through the center of body mass to destroy a vital organ will rapidly lead to death.

As we will see as we discuss different calibers throughout the book, each is typically best suited for a certain range or operational need, for instance the use of .50BMG antimateriel rifles at ranges out to beyond 2,000 meters (2,187 yards) or the use of 5.56x45mm rifles by police snipers to guard against overpenetration. Smaller-caliber rounds are often criticized in the media as lacking the ability to kill an enemy with a single round.

While it is certainly true that the bigger the projectile (and the faster it travels), the greater the chance of incapacitating an enemy with the first round, stories of calibers like the 5.56x45mm being designed only to wound rather than kill and similar tales are myths. Indeed, incapacitating an adversary, as any sniper will advise, is all about shot placement rather than necessarily about caliber.

The FBI agrees:

Shots to the Central Nervous System (CNS) at the level of the cervical spine (neck) or above are the only means to reliably cause immediate incapacitation. In this case, any of the caliber commonly used in law enforcement, regardless of expansion, would suffice for obvious reasons. Other than shots to the CNS, the most reliable means for affecting rapid incapacitation is by placing shots to large vital organs thus causing rapid blood loss. Simply stated, shot placement is the most critical component to achieving either method of incapacitation.

Smaller-caliber rounds actually often have an advantage over their larger cousins. The 5.56x45mm, commonly employed in the M4 and M16 series of rifles, typically yaws when it hits a target. Yawing is the action of the bullet turning upon itself as it travels through the body. This will create a larger permanent wound cavity and helps to offset the smaller caliber. Yawing is more common with lighter and smaller caliber rounds as they are inherently more unstable in flight.

TRAINING THE MODERN SNIPER

Today's sniper is no longer selected simply for being the best shot in the platoon, given a scoped rifle and told to get on with it. The selection and training of today's snipers is second to none in ensuring the right soldier with the right physical and psychological attributes is given every opportunity to excel in his craft. There is a high level of commonality across all sniper schools, particularly in NATO countries, and the basic programs look remarkably similar. One reason for this is that when establishing a sniper school, many nations will look for guidance to premier sniping training programs like the U.S. Marine Scout Snipers and effectively model their local programs on how the Marines do it.

Marine Corps Scout Sniper Platoons run two-week-long screening programs at each Marine battalion to pre-select candidates. During the screening program, Marines must blitz the Corps Physical Fitness and Combat Fitness Tests and confirm Expert on the issue rifle. Candidates that pass are invited to attempt the Scout Sniper Course at one of four locations across the United States (a Scout Snipers Team Leaders Course is also available for suitably qualified personnel).

The Scout Sniper Course itself is currently 12 and a half weeks in duration, including nine weeks of marksmanship, with the remaining time spent on fieldcraft, observation, land navigation and stalking. Prior to 2010, the course was four weeks shorter and had a greater focus on fieldcraft, while today's program, tempered by long experience in Afghanistan and Iraq, focuses primarily on shooting. This is where, in the curious vernacular of the Marine Scout Sniper community, students attempt to transition from Professionally-Instructed-Gunmen (PIGs) to Hunters-Of-Gunmen (HOGs).

The first phase begins the marksmanship portion of the course with the KD or Known Distance range, engaging both static and moving targets at known distances from 300 yards (274 meters) out to 1,000 yards (914 meters). Along with marksmanship principles and instruction, the students gain an understanding of the effects of wind and barometric pressure on long-range accuracy. The student snipers are also trained in observational techniques using rifles optics, binoculars and spotting scopes, and on accurately sketching what they see, an important reconnaissance skill.

Stalking is next and the class set out to move unobserved several hundred meters while instructors search for them with binoculars. The student snipers begin as far as 1,200 yards (1,097 meters) away and must stalk undetected toward their target to close within 200 yards (182 meters) within three hours. If an instructor suspects they have identified a student, a *walker* is dispatched with a radio to guide him to the spot. If the student is discovered, he automatically fails the course.

Once a student has stalked within 200 yards (182 meters), they fire a single shot at a fixed target. The walker is again dispatched to confirm that the student has a line of sight on the instructors. The instructors raise cards with letters printed on them that the student must correctly read to the walker. Once this is completed, a second shot is taken. At this point, the walker checks the student's rifle is correctly dialed in and that the student is positioned in a stable shooting position. Once confirmed, the student has passed the stalking phase.

Either immediately following or immediately before the stalking phase, the students begin the UD or Unknown Distance shooting portion of the course. Here targets will be arrayed at unknown distances. The student and his spotter must use all of their newly taught skills to accurately estimate the range to the target, dial in their rifle optic and engage the target with solid hits, all under a stop watch.

The final phase teaches advanced fieldcraft and reconnaissance techniques. It finishes with an endurance test where groups of 16 students have to evacuate four 200-pound (90-kilogram) dummies simulating wounded Marines over a distance of some 23 miles (37 kilometers) at night. Students who pass this final phase are awarded the Hog's Tooth, a 7.62x51mm bullet attached to paracord to wear around their necks denoting them as HOGs. Some Marines later graduate to the Marine Special Operations Forces Advanced Sniper Course.

The British too run what are known as Sniper Cadres within each battalion to pre-select candidates to attend the sniper course at the Support Weapons School at the Land Warfare Center in Wiltshire. At the height of Operation *Herrick* in Afghanistan, the British Army was training 120 snipers annually. The conflicts in Iraq and Afghanistan have seen the number of snipers in the British military, including the Royal Marines, double within the decade following 9/11. The British course currently lasts for nine weeks. Only one in four candidates passes.

The Australian Army runs sniper courses at the battalion level. Two advanced programs, the six-week Sniper Team Leader and Sniper Supervisor, are run at the School of Infantry, Singleton. "The first part of the basic course is the range module (which) goes for around one week," explained former Australian Army Sniper Master Nathan Vinson: "after that there is a navigation package and then followed by stalking, camouflage and concealment exercises, observation and deduction exercises and finally, judging distance exercises."

Vinson knows sniping after 24 years in the Australian Army and having served as both an instructor and as the Sniper Master at the Australian School of Infantry. Along with serving with Australian forces on Operation *Slipper* in Afghanistan, Vinson conducted a six-month exchange with the U.S. Marines' Scout Snipers to assist in their Afghanistan predeployment training.

The four-week basic program includes shooting at known and unknown distances from between 200 and 800 meters (218 to 874 yards). The sniper program for Australian Special Forces units such as 2 Commando Regiment and the SASR (Special Air Service Regiment) has students shooting out to 900 meters (984 yards) to pass. "Target sizes vary from 200m (Hun's Head) out to full sized targets from 600m to 800m. There is also a requirement to hit moving targets at each range, including the Hun's Head targets," explained Vinson. The Hun's Head, as the name suggests, is a target of a human head clad in a helmet reminiscent of a German helmet of World War II.

The program attracts a typical pass rate of around 20 percent. The stalking phase involves the prospective sniper moving unobserved from a distance of 1½ kilometers (1,640 yards) to within 200 meters

(218 yards) of a target within a time limit of two hours. Like the Scout Sniper course, the Australians are permitted two shots with blank ammunition if they remain unseen. Many students use water from their Camelbaks to dampen the ground and minimize the telltale dust kicked up by the shots.

"The stalk is the culminating point where all of the sniper skills have got to come together to succeed," explained a sniper instructor from 2 Commando Regiment. "This is where the sniper must demonstrate all of his core skills, from navigation, field craft and sense of the terrain. Only this will allow the sniper a silent move to his firing position without being detected."

"The snipers are under constant surveillance and if at any stage they get observed in their movement, or in the preparation of their firing position, they fail and don't progress in the assessment." Again, like the Marine program, the instructors hold up cards with numbers written on them. If the students can read the numbers, they are allowed to progress to the final test: two live fire targets that must be hit with a single round each to pass the course and gain the coveted sniper qualification. The highly sought after Sniper Team Leader course is also offered to qualified snipers to understand the tactical employment of snipers. "As far as I know we are one of the only countries that runs a Sniper Team Leader course, which provides advanced training in navigation, live fire field firing, engagement of multiple targets, firing through glass, mission planning, conduct of orders and multiple team engagements," added Vinson.

The U.S. Army's Sniper School at Fort Benning currently runs a seven-week program known as the Basic Sniper Course. Like the Marine Scout Sniper program, candidates receive instruction in advanced marksmanship techniques on both known and unknown distance engagements, but the course has a greater focus on fieldcraft and the stalk. In 2001 the Army's Sniper School precipitously added an urban combat or more properly MOUT (Military Operations in Urban Terrain) component to the course.

The Sniper School's objective is to "train individuals to perform Sniper missions in a combat environment to include: precision fires on enemy personnel and equipment, intelligence gathering, countersniper operations, infiltration and overwatch of NAIs [Named Areas of Interest], occupation of and operations in support by fire positions, ballistic interdiction of IEDs, and disruption of enemy operations." The course is extremely popular and prospective students often have to wait several years for a slot to open up, even though the school has tripled the amount of places (to some 570) in the ten years since the start of Operation *Iraqi Freedom* (OIF).

During the course, candidates are instructed in:

fieldcraft skills; advanced camouflage techniques, concealed movement, target detection, range estimation, and terrain utilization (Macro and Micro), intelligence preparation of the battlefield (IPB), relevant reporting procedures, sniper tactics, advanced marksmanship; known and unknown distance firing, at stationary and moving targets during daylight and limited visibility in varying weather conditions, and staff subjects (intelligence, mission, training, combat orders, command and control, and training management).

The first week immerses the candidate in the principles of stalking and multiple methods of estimating engagement ranges along with observation and surveillance techniques. The second week sees the prospective sniper start shooting, currently with the semiautomatic 7.62x51mm M110 SASS (prior to the last few years, students used the bolt action 7.62x51mm M24 as their primary platform). They also begin to learn ballistic theory and the effects of the environment on bullets.

In the third week, shooting with the M110 continues, with candidates learning how to gather their own ballistic data and construct range cards, along with a written exam that quizzes them on everything they have learnt thus far, including the mathematical equations required to estimate distance. At the end of the first month, they are introduced to the M24's replacement, the .300 Winchester Magnum M2010, and spend a week shooting it and refining their ability to dial in ranges and account for wind using their scope.

The second month of the Basic Sniper Course begins with instruction on the U.S. Army's current issue sidearm, the 9x19mm Beretta M9 (as most candidates will have likely never fired a pistol), and instruction on firing both the M110 and M2010 against moving targets and the unique challenges of firing at night, including the use of infrared laser illuminators. They also spend more time estimating ranges and engaging unknown distance targets.

The sixth week sees the students introduced to the .50BMG M107 SASR for more range work and a component on operating in urban environments and shooting from different positions. "You're not always in the prone here shooting, you're always shooting [from] different positions and when you're deployed to Afghanistan, Iraq or wherever it may be, you're not always going to be in the prone, you might have to use a tripod to shoot with, or shoot off a rock that's not flat or even off a roof top, which is pretty common," explained an instructor.[8]

By the seventh and final week, the instructors are looking for the students to be registering first-round hits at 600 meters (656 yards) nine out of ten shots with the M2010. The course culminates in a field exercise not unlike that in the Marine Scout Sniper program. It involves the candidate using his newly honed fieldcraft to stalk from 1,000 meters (1,094 yards) out to engage a target at 300 meters (329 yards). If he remains unseen by the instructors acting as spotters, he must read out a card held up by the instructor and then return undetected to his start point. Less than half of the candidates pass.

Other NATO nations such as the Dutch run similar courses. The Dutch Marines have their own Sniper Instruction School. Following a two-day pre-selection, candidates are invited to attend a five-week school in marksmanship followed by a fieldcraft and stalking component.

A Dutch instructor commented on the program: "We do short to mid-range until 800 meters. Today we did 900 to 1,000 meters. We also do the UKD or unknown distance shooting." They also focus extensively on operating at night: "Most of our action (in Afghanistan) is happening at night. For infiltration we use the cover of darkness most of the time as an advantage. We try to train our students with the night vision system to see the pros and cons of night shooting."[9]

The Canadians require prospective snipers to already have passed their Basic Recce Patrolman course, as this provides the candidates with a solid grounding in many of the fieldcraft and observational skills that snipers also need. Prospective snipers must also pass a battalion-level Pre Sniper Course. The actual Canadian sniper program, the Basic Sniper Course, is nine weeks long.

"We cover conventional shooting. We cover field firing shooting, which is unknown distance, and the sniper has to judge the distance to the target," confirmed Warrant Officer George Williams in an interview with *National Defense Magazine* in 2004.[10] Along with shooting and stalking, the course focuses heavily on KIM (Keep In Memory) games and similar observational tasks. "We hide twelve objects in an area, and the snipers have to be able to locate the objects and identify what they are. It is important for a sniper, when he is on the battlefield, to remember what he has seen, come back and accurately report."

Once the student graduates from the sniper course, he can attend the six-week Advanced Sniper School that is somewhat similar to the Australian Sniper Team Leader program. "We will teach them how to instruct, how to advise commanders on the use of snipers," explained Williams. The Advanced Sniper School also offers tuition on the .50BMG McMillan TAC-50 and high and low angle shooting. "We will get pretty heavy and thick into the shooting of our .50 caliber sniper rifle where we take into account factors like barometric pressure, shooting uphill, downhill, air temperature, ammunition temperature..." Upon graduation from this advanced program, the sniper is recognized as a Unit Master Sniper within his battalion.

The NATO Basic Sniper Course run in Germany at the old International LRRP School (a particularly well-regarded NATO Long Range Reconnaissance School that taught generations of snipers and scouts the art of stalking and camouflage) covers similar fundamentals over a four-week program, including the ability to prosecute targets out to 800 meters (874 yards) in both known and unknown distance shooting; tactical infiltration and exfiltration; camouflage and concealment techniques; and observation and distance estimation. All of the courses use a final exercise that tests everything the student has learnt over the proceeding weeks. Although difficult these are perhaps not as tough as the Russian "finishing schools."

When the Russians re-established their sniper school in 1999, they selected the top dozen snipers for further training: according to Les Grau these unfortunates were sent to Chechnya for a month-long "live fire exercise." Of that initial dozen, three were killed in action. Still their returned focus paid dividends during the Second Chechen War. Russian snipers still do train somewhat differently from their Western counterparts, particularly in terms of exfiltration from a hide site should they be compromised. As noted earlier, they are typically deployed with a security element tasked with overwatching the sniper hide site. If this security element is compromised or incapacitated, the Russian sniper is trained to use a red signal flare that will trigger a pre-plotted artillery fire mission onto his position.

The Estonians, who have deployed their snipers to both Iraq and Afghanistan, have an initial five-week marksmanship course followed by the actual sniper program that itself lasts for a further three months and includes advanced sniper skills, stalking and fieldcraft along with a reconnaissance component. Once the candidate has completed the marksmanship course, he must volunteer and be accepted in competitive selection for the sniper course. The Estonians also train their snipers extensively in calling in indirect fire using NATO processes during this phase. All of this training culminates in a two-day live-fire exercise bringing together all the candidate has learned. If they fail any part of this final exercise, they fail the sniper course.

The Danish military also operates a very competitive sniper program. Having been caught without a non-SOF sniper capability at the start of their commitment to security operations in southern Afghanistan in 2006, the Danes quickly established a program drawing heavily on their nation's SOF. The course now includes a week-long pre-selection at battalion level designed to ensure the candidate has the right mix of skills and aptitude to become a sniper.

Once he has passed the pre-selection, the candidate will complete an eight-week sniper training program that culminates in a live-fire exercise with similarities to the Estonian program. The candidate must conduct a successful stalk into a target area before attempting five courses of fire at unknown distance targets out to 700 meters (766 yards).

The four-week French sniper course is also similar to the USMC model and trains candidates to shoot accurately at known and unknown distance targets out to 800 meters (874 yards). Once the prospective sniper is qualified, they attend an advanced programme, again lasting four weeks, that teaches the use of GPS and satellite communications equipment and shooting the .50BMG anti-material rifle out to 2,000 meters (2,187 yards). Again, like the Marine Scout Snipers, French snipers are also trained as forward observers able to call and adjust fire from 81mm mortars.

They also run one of the most technologically sophisticated programmes in existence with extensive use of video-based simulators that project an interactive scenario on the walls of the training center. These simulators allow different scenarios based on contemporary operations in Mali and Afghanistan to be programmed in and the snipers' reactions to different rules of engagement can also be tested.

Vinson explained how sniper training differs between countries:

The basic sniper training no matter what country it is conducted in normally follows the same concept of training ... however the expertise of the trainers is where our courses change. I worked for six months with the USMC Scout Sniper Platoons before they deployed to Afghanistan and also observed their Mountain Sniper Course. When it comes to operational experience the U.S. and UK armies win hands down. They have had more time behind the trigger and more active deployments ... some of the USMC sniper team leaders had deployed, as operational snipers, more than four times.

In the world of special operations forces, most devise and maintain their own courses. U.S. Army SOCOM (Special Operations Command) has the seven-week Special Forces Sniper Course (formerly known prior to 2007 as the Special Operations Target Interdiction Course or SOTIC). The SFSC trains Army Special Forces, Delta, Ranger and selected SOCOM partner units including Air Force Special Tactics. Graduates from the SFSC are known as Level I snipers (as are the graduates of the Marine Sniper Instructors Course at Quantico). Those who graduate from any other program, including courses run at individual Special Forces Groups or Ranger Battalions, are termed Level II snipers.

The SFSC includes components from the Special Forces Advanced Reconnaissance, Target Analysis and Exploitation Techniques Course (SFARTAETC) and an Assault Operations component to prepare candidates to provide precision fire, support during Direct Action missions. The SFARTAETC component trains students in all facets of reconnaissance and surveillance in both urban and rural environments. They even complete a climbing course to enable them to access elevated hide sites. Among the heavy emphasis on the reconnaissance aspect, the students are intensively trained on the M2010, M110 and M107 and are given a familiarization shoot from a helicopter so they can understand the unique requirements of aerial shooting.

The U.S. Army Ranger Regiment formerly trained its snipers through the Basic Sniper Course and, on occasion, the U.S. Marine Scout Sniper program. A select few also attended the SOTIC or SFSC as it is now known. Today, Ranger snipers are given access to a wide variety of specialist sniper courses, both within the military and commercially run, including High Angle shooting programs. Rangers are one of the few units within SOCOM that have full-time snipers. They often receive additional training by Delta's sniper veterans.

Conversely, Navy SEALs have their own program. The Naval Special Warfare sniper program is some three months in length and includes traditional stalking and fieldcraft along with specialist communications, photography and reconnaissance modules. A prospective SEAL sniper is faced with both static and moving targets out to a kilometer in range. SEAL snipers are also taught to operate as so-called "singletons" rather than in pairs as many of their roles, such as aerial sniping or supporting assaulters breaching an objective, do not necessarily require a spotter.

The SEAL Sniper Course includes modules on photography, using encrypted satellite communications systems and a four-week-long stalking component before seven weeks are spent on marksmanship, teaching the students to hit moving targets the size of a human head at 800 yards (732 meters). During much of the 2000s, the course trained individuals with the 7.62x51mm Mk11, the 5.56x45mm Mk12, the .300 Winchester Magnum Mk13 and the .50BMG Mk15. The 7.62x51mm Mk17 and Mk20 have now been added to the program. SEAL Team 6 snipers receive continual further training at all manner of commercial and military schools along with attending allied programs. Their snipers, known as Black Team, are trained so extensively that they will guarantee a headshot, at night, on a target out to 300 meters (328 yards).

Delta Force snipers are part of the Recce Sniper Troop in each squadron and are invariably older and more experienced operators with five to six years in the unit before they become Recce snipers. Delta snipers will attend SFSC along with any number of high-end civilian schools and many attend the British SAS sniper course (details of which are a closely held secret – as is customary with UK Special Forces training). They also receive advanced mountaineering and climbing instruction along with survival skills, as they are expected to operate alone in any environment on Earth for as long as the mission requires.

Snipers must practice shooting at high angles, as the greater the angle they are firing at, the greater the loss of velocity and thus eventual loss of range. The projectile will also shoot low from the point of aim due to the increased drag experienced on the round. Shooting downhill also requires practice, as this will increase velocity and shift the impact point above the point of aim. Many snipers learn to use the Angle Cosine Indicator, a device that is attached to the rifle and compensates for the angle of the rifle for final scope adjustments.

Along with shooting at altitude, snipers today increasingly find themselves operating within urban environments. As we have noted, the U.S. Army's Basic Sniper Course now includes an urban module that teaches candidates how to construct and operate from an urban hide site. The USMC has instituted a 16-day Urban Sniper Course that somewhat incongruously includes components on maritime shooting along with aerial shooting from a helicopter. The most extensive urban sniping training available in the U.S. is, however, as part of the Special Forces Sniper Course.

NATO-sponsored programs, such as the International Special Training Centre's (ISTC) High Angle/Urban Sniper Course in Germany, provide the opportunity for both conventional and SOF snipers to participate in a multinational program that combines instruction in both disciplines. The course includes everything from stalking and zeroing weapons in alpine and mountainous environments to shooting across valleys out to 1,800 meters. One essential component to prepare for deployment to locations such as eastern Afghanistan is the instruction in shooting up and down slope at angles of up to 40 degrees. "This course not only strengthens the alliances between attending multinational personnel, but expands the capabilities of our partner special operations forces," stated the ISTC's director.[11]

Due to the flat terrain in their home country, Dutch snipers even conduct an annual training package at the Mountain Snipers Course at Marine Corps Mountain Warfare Training Center in Bridgeport, California. This prepared them for operations in Afghanistan by allowing them to practice high-angle shooting. Other nations also attend the Marine Course, including British Royal Marines.

Fort Benning's Infantry School also runs a Mountain Warfare Course that angles altitude and high-angle shooting. "We are training snipers and our other marksmen to engage not only targets at greater ranges, but also those above and below them, where a rifleman may overshoot or undershoot. Even with the latest and best optics, range estimation becomes even more critical in the long ranges of the mountain fight," explained Major General Walter Wojdakowski, commander of the school.[12]

As briefly mentioned earlier, the use of snipers as mortar forward observers and sometimes as forward air controllers has increased in many armies. For the later task, a JTAC (Joint Tactical Air Controller) will often be attached to the sniper team to be able to direct aerial fires (or an MFC or Mortar Fire Controller to direct mortar assets). Some sniper teams however receive basic instruction in calling in air support, while many SOF snipers, such as those of the British SAS, receive much more comprehensive training which is closer to that received by actual JTACs. For the former, many sniper team leaders will be qualified in mortar fire control: "The Team Leaders Course has a Call for Fire package, however you will find that most infantrymen in the battalions are trained at some level to call in mortar support," said veteran Australian sniper Nathan Vinson.

DESIGNATED MARKSMEN

Up until 1960, the U.S. Army had issued one .30-06 M1D sniper rifle per squad, while the Russians issued a 7.62x54mm Dragunov SVD, firstly at the platoon and later at the squad level. The West German Bundeswehr adopted a similar tactic during the Cold War with one man in each squad armed with the 7.62x51mm Heckler and Koch G3 SG/1, a modified G3 with adjustable stock, folding bipod and optics. All of these were what one would today term Designated Marksmen.

The British Army instituted the practice of arming one man, the best shot in the platoon, with the bolt-action 7.62x51mm L96A1 sniper rifle as the Platoon Marksman. This practice was a direct result of their early experiences in Iraq and Afghanistan where the benefits of an additional long-range weapon with powerful optics soon became clear. The L96A1 was chosen chiefly because of its optics and its availability, as it was being replaced in the sniper sections by the .338 Lapua L115A3. The 5.56x45mm L86A2 Light Support Weapon or LSW, basically a longer-barreled, bipod-equipped version of the standard L85A2 assault rifle, was pressed into service prior to the L96A1s becoming available.

The Platoon Marksman often traveled with the platoon commander in the field to provide both intelligence gathering and positive identification along with precision fire as needed. As the British Army became embroiled in combat operations in Helmand Province, Afghanistan, an urgent request was placed for a semiautomatic platform for the Platoon Marksman. In 2009, stocks of the 7.62x51mm L129A1, a semiautomatic AR-10 style platform, began to be delivered.

The Germans reintroduced the marksman concept in 2010 with the section-based shooter expected to identify and engage targets out to 600 meters. Their experiences in Afghanistan and joint operations with the Americans fostered the reintroduction. In the U.S. Army and Marines, Squad Designated Marksmen were deployed from around 2003 or 2004 with one SDM assigned per squad, often equipped with a scoped 7.62x51mm M14 or even a variant of the 5.56x45mm M16A4.

One of the key duties of the DM or SDM in today's Western militaries is to use their optics to identify and classify targets prior to engagement by the rifle squad. In environments where insurgents will attempt to enter or leave the battlespace disguised as civilians, this task is of critical importance. In urban combat they are also often the first line of defense against insurgent snipers and suicide bombers. Likewise, in the absence of a dedicated sniper team, the DM can prove essential in providing overwatch for his squad and platoon, as one U.S. Army Designated Marksman explained:

> On patrol, especially in a mountainous environment like many parts of Afghanistan, immediate enemy threats can appear at several hundred meters away ... often times we don't have snipers attached to our squad, so having an SDM that can fill that role and maintain his position as a squad member performing traditional rifleman tasks multiplies the capabilities of the squad.[13]

The U.S. Army Ranger Regiment now runs battalion-level training courses for its DMs. Some Rangers, including a number of snipers from the battalion sniper platoons, also attend the Army Marksmanship Unit's two-week DM program, which is known as the Squad Designated Marksman (SDM) Course. The SDM Course covers engaging targets in what the instructors term "no-man's land," the distance between 300 meters (328 yards) and 600 meters (656 yards). Infantry fire teams are expected to be able to engage point targets up to 300 meters (328 yards), while snipers are traditionally shooting to beyond 600 meters (656 yards), although as we shall see in later chapters, the war in Iraq particularly has modified this thinking.

HOW ARE SNIPERS ORGANIZED?

As we noted in the first chapter, the Red Army in World War II were largely credited with the first implementation of the sniper pair, rather than dispatching a solitary sniper to the battlefront. The U.S. Army explains the thinking in their SOTIC manual:

> The sniper pair consists of two equally trained snipers who provide mutual security and support for each other. Snipers are employed in pairs for the purpose of enhancing the team's effectiveness and diminishing the stress that a single sniper would be more apt to encounter. Sniper pairs may also

engage targets more rapidly and may stay in the field for longer periods of time than a single sniper. Additionally, a high priority target may warrant that both snipers engage the target to ensure a hit. The two-man concept permits this flexibility.

Experience in World War I, World War II, etc., has shown that deploying snipers in pairs as a sniper/observer team significantly increases the success rate of the missions. With few exceptions, snipers who are deployed singularly have shown a marked decrease in their effectiveness and performance almost immediately after the start of the mission. This is due to the individual becoming overwhelmed with concern for his security, the tasks to be accomplished, and his own emotions, i.e., fear, loneliness, etc.

U.S. Marine Corps Scout Sniper platoons typically comprise between eight and ten two-man Scout Sniper teams. The Scout Sniper platoon is a battalion-level asset and is typically placed under the battalion Headquarters Company or Weapons Company. Upon deployment, the platoon is often broken up with teams detached to individual companies and platoons wherever they are needed. For specific intelligence-led operations the platoon may deploy as a single entity.

Officially, according to the Marine Corps sniper manual, the duties of these Scout Sniper teams are:

Spotter: Detects, observes and confirms sniper targets, calculates the range and wind conditions on a given target, conducts reconnaissance and surveillance.

Sniper: Delivers long-range precision fires on selected targets, conducts reconnaissance and surveillance.

Each team, or sniper pair, is equipped with one standard-issue sniper rifle, currently the bolt-action 7.62x51mm M40 and their personal M16A4s or M4s (the Corps has recently decided to transition completely from the M16A4 to the M4). Additionally the team may have access to a semiautomatic sniping platform like the 7.62x51mm M110 that can equip the spotter. Four .50BMG M107 SASR platforms are available per Scout Sniper platoon.

In Stryker units within the regular U.S. Army, snipers tend to be organized at the company level with a three-man sniper team of team leader, sniper and spotter. This team is generally equipped with one 7.62x51mm M24 or M2010 sniper rifle, one .50BMG M107 SASR, one 7.62x51mm M110 SASS and personal carbines including at least one mounting a 40mm M203 grenade launcher for providing security to the sniper hide.

In U.S. Army Cavalry units they are organized into a sniper section that sits at battalion level. Each section comprises two three-man sniper teams and the section leader. Soldiers assigned to the sniper section are not necessarily school-trained snipers. They are often selected based on an internal competitive selection from within the battalion. As spaces become available on the always heavily subscribed Basic Sniper Course, soldiers are rotated through.

The British Royal Marines in comparison organize their snipers as part of their Brigade Recce Troop (a troop in Royal Marine terms is equivalent to a platoon). Each Recce Troop can field, at full strength, five two-man sniper pairs. On deployments in Afghanistan, the sniper pairs were parceled out to each individual rifle company to provide each with a sniper capability. The British Army currently fields at full strength (which it rarely is), 16-man sniper platoons at battalion level. They are placed under the Support Company of the battalion along with the reconnaissance platoon and the heavy weapons platoons.

The Australian Army organizes its snipers in an infantry battalion as a sniper cell. "Standard manning for the sniper cell is 9 personnel, so four pairs and a Sniper Supervisor. Some battalions have raised sniper platoons in the past but that is at the sacrifice of other platoons in the battalion, so not a common practice," explained Vinson. In terms of weapons, the Australians have an impressive range available: "SR98 7.62mm bolt action sniper rifle, SR25 semiautomatic 7.62mm rifle, H&K 417 7.62mm semiautomatic weapon, .50 caliber Barrett semiautomatic anti-materiel weapon, and the primary operational sniper weapon, the Blaser .338, with (the option of) 7.62mm barrel," he added.

Estonian snipers (who served alongside the British in Afghanistan) are attached at company level as part of a battle group structure. They operated in two-man pairs although frequently they deployed in multiples. There are 12 snipers in an Estonian sniper section. The Canadians typically deploy operationally in four-man teams of two sniper pairs. They are also organized at battalion level with eight snipers. The French have both light and heavy sniper teams (the Light Team using 7.62x51mm platforms and the Heavy Team employing the .50BMG). Two of each of these types of team made up the sniper group in a French infantry company. Three such groups were available across a French battalion.

In the world of SOF things are always done a little differently. In the so-called Tier One units like Delta Force and SEAL Team 6, their snipers are organized into their reconnaissance troops within each squadron. Each troop numbers somewhere in the region of 16 operators although it can be as few as 12. Each squadron normally comprises four troops, with the Recce Sniper/Black Team (as it is known within Delta and SEAL Team 6 respectively) being one of the four, with the others detailed as Assaulters.[14]

The Rangers organize their snipers into a sniper platoon under each Battalion Headquarters. Prior to the end of the 1990s, they had attached the snipers to the Company Weapons Platoon along with the 60mm mortars and M240B medium machine guns. As noted earlier, Ranger snipers are full-time snipers as the Rangers have battalion-level reconnaissance platoons and the Ranger Reconnaissance Company to handle dedicated reconnaissance tasks. Conversely, Delta and SEAL Team 6 snipers must be both accomplished marksmen and reconnaissance specialists.

Just as important as the physical makeup of the sniper team is how and where they are placed within the command structure. For many years, both after World War II and after Vietnam, snipers were largely forgotten or at best ill-managed, with very little thought going into how or where they were deployed on the battlefield. As we've noted earlier, it's a sad fact that snipers tend to be marginalized if not disbanded when a nation is not at war.

British sniper Sergeant Craig Harrison explained the problem in *The Longest Kill*:

> The challenge that the sniper has is that most officers don't understand how to employ snipers and often don't understand the basics of marksmanship. They want to tell us where to go instead of letting us pick the best position. If an officer doesn't know how to use us, doesn't understand the challenges of shooting at long distances, and then gives us too restrictive ROE [rules of engagement], our job becomes a nightmare.[15]

Traditionally, sniper sections were organized under the weapons company of infantry battalions. While a logical location in some respects, as they add another weapon platform to the support options, it saw them deployed in much the same way as one would deploy light mortars or machine guns. The hugely valuable contribution snipers can make to reconnaissance and intelligence gathering was rarely utilized to its fullest extent. Other snipers saw themselves attached in any number of strange locations on the order of battle: historian Adrian Gilbert quotes a Parachute Regiment officer who explained, with some obvious frustration, that up until the mid-1980s, his snipers were drummers first, snipers second.

In other organizations, snipers were part of a reconnaissance platoon, or in the U.S. Marine Corps, the Surveillance & Target Acquisition platoon. In Australia they were likewise part of each battalion's Recon-Sniper platoon. While this is a more natural fit, and certainly ensured the reconnaissance and surveillance capability inherent in a sniper section was far better employed, some argued that it tended to negate against the sniper's kinetic capabilities: to put it bluntly, they worried that with all the reconnaissance focus, the snipers wouldn't be shooting! To some degree it is a valid concern.

Operationally snipers seem to work best when tasked by the battalion or Task Force commander but logistically located within the intelligence shop or reconnaissance platoon. This tends to ensure they are deployed realistically with at least some understanding of their capabilities. A sniper must still be able to explain how best to employ their specialist capability, however, as they may often be attached to other units for a particular operation.

A Special Forces sniper instructor interviewed by author Gina Cavallaro explained:

> You can't take for granted that your commander knows your capabilities or when to use you. A sniper needs to understand his own capabilities and limitations because he needs to be able to brief that to his commander and brief it well. The sniper has to be able to articulate to his commander in favour of and against a mission … So if you can make the commander understand you've got the capabilities to perform a [sniper] job and you can sell him on that job, then you'll be put into action and used much more as a sniper.

Another Special Forces officer agreed: "Snipers are like more systems in the Army. Every commander has them, but only about 5 percent know how to effectively use them."[16]

SNIPER COMPETITIONS

Every year Fort Benning hosts the International Sniper Competition. According to the U.S. Army:

> Sniper Teams from across the globe will travel to Fort Benning to compete in the Annual International Sniper Competition. The goal of this competition will be to identify the best sniper team from a wide range of

agencies and organizations that includes the U.S. Military, International Militaries, and Local, State and Federal Law Enforcement.

The best team will be identified only after all teams have completed a gauntlet of rigorous physical, mental and endurance events that test the range of sniper skills that include, but are not limited to, long-range marksmanship, observation, reconnaissance and reporting abilities, and abilities to move with stealth and concealment.

Past winners have included teams from the Canadian Army (2001), the U.S. Army's 501 Parachute Infantry Regiment (2002 and 2004), the U.S. Army National Guard's 19th Special Forces Group (2003), the National Guard Sniper School (2005), the 75th Ranger Regiment (2006), the U.S. Army Marksmanship Unit (2007, 2008, 2012 and 2013), the U.S. Marine Corps Scout Sniper School (2009), the U.S. Army's 1st Special Warfare Training Group (2010), the U.S. Army 3rd Special Forces Group (2011), the U.S. Army's 1st Special Forces Group (2014) and the Irish Ranger Wing in 2015.

Fort Bragg also hosts the U.S. Army Special Operations Command Sniper Competition, which in 2015 saw over 20 international teams compete for a grueling four days with shoots ranging from 50 to 1,000 meters. Denmark's snipers likewise conduct an annual contest that has at least one rather unique event, sniper golf. This requires the sniper team to hit a golf ball swinging in the wind from a foot-long tether, placed at unknown distance ranges out to 250 meters.

Like the various international sniper competitions, the Australian SASR also run what is known as the annual Sniper Concentration. Snipers from all of the Army battalions are invited to attend for a competition shoot and an invaluable training symposium with SASR sniper instructors. It also fosters an informal information exchange between the Special Forces snipers and their infantry counterparts. Participants typically shoot over 600 rounds during the seven days, far more than they are allotted at their battalions. They also are given instruction on Special Forces specific sniping platforms and observation devices.

An SASR Corporal commented on the 2015 program: "These competitors are all competent snipers, having passed their basic sniper course and some have gone on to do their team leaders course. Now's the time for them to step up and demonstrate their proficiency and experience within their specialist skill set. There's nowhere to hide; at the end of the two weeks the scores will tell the story." In that particular year, SASR won overall, followed by pairs of regular infantry snipers from First Battalion, Royal Australian Regiment (1RAR) and Seventh Battalion, Royal Australian Regiment (7RAR).

SNIPER PSYCHOLOGY AND PTSD

The wars in Afghanistan and Iraq have brought the prospect of psychological casualties to the fore, including what is known as Post Traumatic Stress Disorder or PTSD. Military charity Combat Stress defines PTSD as "essentially a memory filing error. It can happen when people are exposed to an extraordinary life-threatening situation which is perceived with intense fear, horror and helplessness."[17] The soldier's training thankfully takes over in such a situation and allows them to fight through the fear. After the immediate danger has passed the mind will attempt to file away the traumatic experience.

> The problem is that when the mind presents the memory for filing it can be very distressing. The mind repeatedly and automatically presents the memory in the form of nightmares, flashbacks and intrusive unwanted memories. These "re-experiencing" phenomena are the mind's way of trying to file away the distressing memory. The re-experiencing can be very unpleasant and distressing because of the nature of the traumatic experience it exposes the sufferer to.

Snipers are not immune from the often debilitating effects of PTSD. Some research, however, suggests that due to their training and the nature of their role they may be better equipped than the average soldier to deal with some of the stresses and triggers that lead to PTSD. According to preliminary research carried out by the Canadian Department of Military Leadership and Psychology, Canadian snipers are apparently coping better with the stresses of combat than their infantry colleagues. The study looked at 19 snipers who had served in Afghanistan.

These snipers also rated their most troubling combat experiences, which makes fascinating if harrowing reading:

1. Knowing someone seriously injured or killed
2. Having a member of your own unit become a casualty
3. Receiving incoming artillery, rocket or mortar fire
4. Improvised IED or booby trap exploded near you
5. Being in threatening situations when you were unable to respond because of ROEs
6. Seeing dead or seriously wounded Canadians
7. Seeing ill or injured people you were unable to help
8. Working in areas that were mined or had IEDs
9. Had a close call or dud land near you
10. Seeing dead bodies or human remains

The Canadian report also neatly summarized the academic work on the psychological impact of killing:

By most accounts killing is a squalid business, so an important question is: what impact does this work have on those who do the killing? Some research-based works by Dyer (2005), Bourke (1999), Grossman (1995, 2004), MacNair (2002) and others suggest that many people who have killed in combat experience some level of remorse and then need to spend time rationalizing and accepting their actions. Whether this remorse necessarily leads to post traumatic stress disorder or other serious mental health issues is not certain, however, because research has shown that not everyone who has killed in combat will suffer social-emotional consequences (Hendin & Haas, 1984a, 1984b).

Some people cope better than others, perhaps because of certain personal characteristics like hardiness or resilience. It is also possible that more dire outcomes are experienced by soldiers engaged in close-quarter, face-to-face killing, leaving long-distance shooters like snipers, artillery men and women, and pilots less prone to suffer much remorse or guilt from their combat experiences.[18]

Much of the academic thought related to the effects of killing has been informed by the work of former U.S. Army Lieutenant Colonel Dave Grossman. He has studied a large amount of anecdotal evidence principally from the Vietnam era to support his findings. He outlines what he believes are three stages of psychological response to killing. The immediate psychological response to killing is often one of euphoria or satisfaction. This is followed, at least in some soldiers, by remorse at the act or the loss of a human life. Finally the soldier will attempt to rationalize the killing to be able to psychologically deal with the act. It is important to note that not all soldiers go through all three stages and some may experience none.

Another Israeli study of 31 snipers in 2005 found that:

The authors noted that the snipers held conflicting attitudes vis-a-vis their enemy, dehumanizing them somewhat while simultaneously recognizing their humanity. Many of the snipers felt remorse and regret at having killed enemy combatants, but they also felt justified, particularly in those cases where their target was engaged in hostile action against Israeli forces.[19]

U.S. Marine sniper Chuck Mawhinney reports that he still dreams of a Vietnamese soldier, but not one of his kills. Instead it is a man he missed. Mawhinney is haunted by the possibility that the enemy he missed may have killed other U.S. service members. Famously, former SEAL Chris

Kyle, the author of *American Sniper*, suffered from classic PTSD from his multiple deployments. Others view operations as an extension to all of the training a sniper must endure and can compartmentalize their experiences, at least for a time (PTSD can commonly appear a number of years after the trauma).

A British sniper in Basra, southern Iraq, in 2003 summed up his attitude toward his job as follows: "People might think we are a bit crazy, but we just think of it as our job." He explained, "There are mortars and rounds coming in, but other than that, it's just like an exercise. It's the first time I have been at war, but we've got a good weapon with a very accurate sight, so it's fine." Another British sniper agreed: "Thinking about it afterwards," he said, "I don't feel much. It's a job in the end. It's either them, or one of us."[20]

On the flip side of the coin is decorated former sniper, Sergeant Craig Harrison. We will discuss Harrison's accomplishments in a later chapter, but Harrison's story is also a salutary lesson on the effects serving as a sniper can inflict upon an individual. Harrison was interviewed for a media release that mistakenly included his name, an uncensored image of his face and details of where he lived and his family. Apparently someone within the Ministry of Defence lacked the common sense to apply even basic PERSEC (personnel security) measures to the article, measures that are normally applied to servicemen and women in roles at risk from terrorist retribution such as Apache aircrew, Special Forces and snipers.

The sniper and his family began to receive threats and a plot was uncovered to kidnap and execute him (a suspect vehicle was recovered by police containing a photograph of Harrison and, ominously, plastic sheeting in the boot). Harrison and his family were appallingly treated by the MOD and the Army, which did nothing to improve the effects of PTSD and a traumatic brain injury Harrison had suffered. Thankfully he left the Army, sued and won damages from the MOD for revealing his identity, took on a role with Accuracy International and began to rebuild his life.

CHAPTER 3
SNIPING IN AFGHANISTAN

SNIPING IN OPERATION *ENDURING FREEDOM* (OEF)

The conflict in modern Afghanistan can be separated into several phases. The first is the downfall of the Taliban in October to December 2001. This was primarily an SOF war by proxy as the Afghan Northern Alliance did much of the fighting on the ground, aided and abetted by U.S. airpower. The second phase saw the United Nations' International Security Assistance Force (ISAF) attempt to stabilize the war-torn country. The third phase resulted from ISAF's expansion into the south and the reemergence of the Taliban as an active insurgency. We will consider the role of sniping in all three phases before we look at specific types of operation conducted by Coalition snipers in Afghanistan.

As we've noted, those early days of the war in Afghanistan were largely a Special Forces fight. The U.S. Green Berets, along with CIA paramilitaries, recruited and fought alongside the Afghans of the anti-Taliban Northern Alliance. It was mainly a war of laser-guided airstrikes, not a sniper's war. The Green Beret Weapon Sergeants brought their sniper rifles with them but the majority of the destruction they wrought was from the air. In the first few months of the war, it was actually rare to encounter an ODA (U.S. Army Special Forces Operational Detachment-Alpha) that had fired their own small arms in anger. This soon changed as the war progressed into the mountains bordering Pakistan.

Usama bin Laden had been tracked into the mountains of Tora Bora, a former base area for the mujahideen in their war against the Soviet Union. Bin Laden knew Tora Bora well from those days, and as his Taliban hosts surrendered and key cities fell to the combined might of American airpower and the Northern Alliance, he retreated there once more with his surviving followers, a motley band of *Afghan Arabs*, many veterans of the earlier war. The term Afghan Arabs described foreign jihadists who had traveled to Afghanistan to fight alongside firstly the mujahideen in the 1980s and later the Taliban in the Afghan Civil War of the 1990s. They became the core of bin Laden's al Qaeda.

A small force of Americans including Green Beret and Delta Force snipers were dispatched to the mountains on the trail of bin Laden in the mountains of Tora Bora. Intelligence had indicated that a wounded bin Laden was sheltering in the cave systems and Delta were brought in to hunt him down. Working with unreliable, and arguably duplicitous, Afghan allies whose loyalty had been bought by the CIA, and with a risk-averse chain of command, Delta was hamstrung, although they took the initiative and pushed the operational envelope whenever they could.

Recce Sniper operators scaled a sheer cliff to establish an observation post, allowing them to call in airstrikes on the entrenched al Qaeda defenders. A similar small team of Recce snipers went after bin Laden himself, triangulating on his radio signals, but were confronted by a mercenary Afghan militia who turned their guns on the Delta operators. Frustratingly none of the special operators caught a glimpse of bin Laden through the sniper optics on their heavily customized 7.62x51mm SR-25s.

During 2001 and into 2002, an elite organization worked alongside the CIA and Special Forces known as Advanced Force Operations or

AFO. Largely comprising Delta's B Squadron Recce snipers, officially its role was to conduct *intelligence preparation of the battlefield* or in the words of CENTCOM (U.S. Central Command) commander General Tommy Franks: "Get some men out into the frontier to figure out what's going on. Find the enemy, then kill or capture 'em." Part of that frontier was the Shahikot Valley in eastern Afghanistan, where al Qaeda and other foreign fighter elements including the infamous Afghan Arabs had taken refuge from the pummeling effects of U.S. and Coalition airpower.

AFO dispatched teams of Recce snipers to conduct an environmental reconnaissance of the area with the aim of understanding, and experiencing, where the enemy were operating. These teams infiltrated on specially muffled All-Terrain-Vehicles with infrared headlights and on foot, with one team braving a snowstorm and a perilous climb to reach their hide site deep in the mountains. The information the Delta operators brought back was invaluable in planning what would become Operation *Anaconda*.

Anaconda was planned as an air assault by helicopter, inserting soldiers of the 10th Mountain and 101st Airborne Division into the valley to flush out the insurgents who were believed to be living on the valley floor. Supporting this effort was a Special Forces-led force of locally recruited Afghan militia called Task Force Hammer who, it was planned, would enter the valley from one end to flush the insurgents out and into the guns of Task Force Anvil – the 10th Mountain and the Rakkasans of the 101st Airborne who would establish blocking positions on the valley floor. Supporting the operation were three AFO teams, two from Delta and one from SEAL Team 6's Recce snipers, along with a number of Coalition SOF teams who had established covert observation posts high in the mountains to block off possible escape routes.

The operation went according to plan until an AC-130 Spectre fixed-wing gunship mistakenly fired upon the joint Green Beret and Afghan force as they approached the entrance to the valley. This mistake, along with a lack of planned preparatory airstrikes and coming under uncharacteristically accurate insurgent mortar fire, convinced the Afghans that discretion was the better part of valor and the advance stalled. Despite being informed by AFO that the insurgents were living in caves on the steep sides of the valley, not down on the valley floor as expected, the air assault went ahead, with only luck stopping one of the heavily laden Chinooks from being shot down.

The AFO Recce sniper teams vectored in continual airstrikes from Coalition aircraft stacked over the battlespace. The third Recce sniper team, a SEAL Team 6 element, discovered an al Qaeda 12.7mm DShK heavy machine gun post occupying their planned OP. The SEALs coordinated with the AC-130 overhead and opened fire on the insurgents before bringing the deadly aerial fires of the Spectre to bear. The same sniper team managed to later eliminate an insurgent force that was attempting to encircle a Task Force Anvil headquarters element with precision fire from their suppressed 7.62x51mm SR-25s and 5.56x45mm Recce rifles.

Rakkasan snipers on the valley floor had their work cut out for them. Equipped with the 7.62x51mm M24, they were at a disadvantage shooting uphill, losing velocity and range. Additionally, the crosswinds played havoc with their rounds. One sniper team responding to insurgent sniper fire used a nearby M240B machine gun team to judge the wind effects: "I asked the 240 guys to give me a three-round burst so I could see the behavior of the tracer. That wind took it for a ride."[1] Waiting until the wind seemed to be at its lowest ebb, the sniper fired. His first round impacted 3 inches to the right of the insurgent. Cycling the bolt, he lined up a second shot and hit the insurgent's center of body mass. The laser rangefinder read just under 750 meters (820 yards).

As the assault in the valley continued, politics intervened at the AFO and SEALs were brought in to relieve and replace the three AFO teams already deployed in the Shahikot. Despite the strong objections of the Delta commander, the three SEAL teams, only one of which was a Black Team sniper element and thus trained in mountain warfare, were infiltrated onto the peaks ringing the valley. The arrival of the SEALs was an example of the inter-service rivalries and jealousies that plagued U.S. SOF units at the time – commanders wanted the SEALs "blooded" and seemingly were less concerned whether they were correctly trained and equipped for success.

The one Black Team element of seasoned Recce snipers, callsign Mako 30, was inserted onto the crest of a mountain named Takur Ghar. As their Chinook began its descent, they were caught in a vicious ambush from concealed insurgents firing RPGs and machine guns. As the Chinook gained altitude to escape the deadly fire, one of the SEALs, preparing to leap from the aircraft moments before, fell out from the open ramp and landed in the snow of Takur Ghar. Surrounded by insurgents, he activated his emergency infrared strobe and began to fight for his life.

Mako 30's Chinook had been badly damaged, but the pilots managed to land the helicopter safely nearby. A second Chinook arrived to ferry the SEALs back to Takur Ghar to attempt a rescue of their comrade. The arrival of this second helicopter on the peak was a surprise to the insurgents and Mako 30 managed to land before the enemy realized what was happening. The SEALs were soon in a pitched battle, however, and were forced to withdraw with several wounded.

An Army Ranger platoon was deployed at nearby Bagram Airbase as a Quick Reaction Force. Summoned to Takur Ghar to assist the SEALs, in the confused environment no one told them not to land on the peak. As they did so, their Chinook was shot out from under them, crash-landing on the mountain. The Rangers were trapped on the peak by insurgents located in hidden bunkers until they were reinforced by a second Ranger squad and, after a number of "Danger Close" airstrikes, stormed and captured the defensive positions in a textbook platoon-level attack. The debacle of Takur Ghar claimed seven American lives.

During operations supporting *Anaconda*, a small group of Canadian snipers broke a world record in April 2002, and then a month later broke it again. Master Corporal Arron Perry made a 2,526-yard (2,310-meter) shot with his bolt action .50BMG McMillan TAC-50, beating Carlos

Hathcock's record by 157 yards (144 meters) and for a short time entering the record books. A month later, his sniping partner Corporal Rob Furlong beat his record with a 2,657-yard (2,430-meter) shot of his own.

Furlong was initially deployed with his sniper team at a height in excess of 10,000 feet. The altitude means the air is thinner, affecting both the velocity of the bullet (it travels faster as there is less resistance) and also flight time to the target, which is consequently reduced. Indeed a new set of equations for the scope is required every 1,000 feet. He managed to hit an insurgent at a very respectable 1,640 yards (1,500 meters) while compensating for winds from three directions over the path of the bullet.

He then spotted a three-man team of insurgents, one carrying a 7.62x39mm RPK light machine gun, moving across a distant ridge in an effort to reinforce the fighters further down the valley. His spotter lased the target and reported a range of 2,657 yards (2,430 meters). The altitude for this shot was estimated to be 8,500 feet. Furlong had taken to heating up his ammunition in the sunshine in an attempt to increase the burn rate of the powder and perhaps make a marginal increase to the range of the round. He would need every last inch he could wring out of the .50BMG.

Once the spotter talked Furlong onto the target, the sniper decided to use the *ambush* or *trap* method and plotted his crosshairs on a point the insurgent would have to cross. After having maxed out the elevation and windage drums on his scope, he used the mildots in his optic to judge the elevation and compensated a total of 4 mils or approximately 15 feet up and to the left. Ensuring as optimum a cheek weld as possible, Furlong triggered the first round.

The round missed, but the insurgents realized someone was shooting at them. Surprisingly they stood stationary, trying to judge where the bullet had come from. Furlong's spotter provided adjustments based on his read of the swirl or vapor trail of the bullet (this is only visible during very extended range shooting) and the splash of the missed round striking the rocks. Furlong adjusted his hold-off and triggered a second round that struck one of the insurgents' backpacks! Knowing he was on target, Furlong rapidly worked the bolt and squeezed off a third round.

The sniper waited with bated breath for the approximately four seconds it took for the round to reach the target. The spotter reported a hit and the RPK gunner collapsed with a massive chest wound. The shot was all the more impressive as Furlong at the time reportedly had no specialist training in shooting at altitude. He was also shooting U.S. issue .50BMG match ammunition that he was unfamiliar with after the sniper teams had expended all of their Canadian issue ammunition. Furlong and the Canadian snipers were officially credited with 20 kills during Operation *Anaconda*.

For a number of years after *Anaconda*, Afghanistan became something of a forgotten war as the United States and Coalition forces invaded Iraq, taking many of the resources necessary to secure the peace in Afghanistan and deploying them to the Middle East. Afghanistan's former rulers, Mullah Omar's Taliban, took the opportunity to rebuild their forces and began to infiltrate back into Afghanistan from their base areas in Pakistan. By the time ISAF decided to attempt to push out security operations into the restive south of the country, the so-called third phase, the Taliban were preparing for an insurgency.

SNIPING IN THE AFGHAN COUNTERINSURGENCY

"I think we are the best thing a commander can have in his hands. We are the eyes and ears of the commander," said one Dutch sniper in Helmand. "We can give precision fire at long ranges and be support for all the troops by going forward to scout the enemy."[2]

The counterinsurgency war arguably began in Afghanistan in 2006. The British agreed to operations in Helmand Province while the Canadians took Kandahar. Both deployments stirred up a proverbial hornet's nest of warring tribes, Taliban insurgents, foreign fighters and drug militias, all of whom had their own reasons to fear and resist the encroachment of a central Afghan government into their provinces. Although the author has termed this a counterinsurgency war, veterans will often remark that Afghanistan was the closest to conventional warfighting they had experienced since the initial invasion of Iraq. Certainly the numbers support the theory: for instance, half a million rounds of small-arms ammunition were expended in combat by the first British battle group into Helmand Province in 2006.

By 2006, Afghanistan had become very much an infantryman's, and a sniper's, war. The enemy could not be routed out and destroyed by airpower alone. The counterinsurgency doctrine had become one of "clear and hold" – clear the insurgents in a particular area by offensive operations before using local security forces of the Afghan National Army (ANA) and Afghan National Police (ANP) to make good on these gains. Unfortunately, while the Coalition forces were largely successful with the former, the second part of the equation often failed thanks to too few troops and the inconsistent quality and morale of the Afghan forces.

Coalition forces faced an enemy that was skilled in the use of the dreaded Improvised Explosive Device or IED that, as the war progressed, became the number one killer of Western soldiers.

The U.S. Department of Defense defines an IED as "a device placed or fabricated in an improvised manner incorporating destructive, lethal, noxious, pyrotechnic, or incendiary chemicals and designed to destroy, incapacitate, harass, or distract. It may incorporate military stores, but is normally devised from non-military components."

These IEDs would be triggered in a number of ways. The pressure-plate or victim-operated IED (PPIED or VOIED) was a simple device where the weight of a passing soldier (or civilian) would detonate a charge of homemade explosive. The second type, the remote-control IED (RCIED), was far more common in Iraq but surfaced on occasion in Afghanistan. This simply gave the insurgents the capability to detonate their devices from a distance by using a mobile (cell) phone or repurposed remote control device like a garage door opener. The third type was the similar command-wire IED (CWIED), although this required a physical connection between the trigger-man and the bomb. This was very common in Afghanistan.

Although there were numerous technological and intelligence-driven countermeasures developed in the war against the IED, the sniper played a prominent role in defeating the devices and the men who built and emplaced them. We will cover a number of operations as examples later in this chapter, but put simply, the sniper could be deployed as either an active countermeasure (for instance in sniper-initiated ambushes of IED emplacers) or a reactive one (engaging insurgents suspected of emplacing or detonating an IED). They were also used in support of direct-action raids against IED cells, lethally targeting individuals identified as bomb makers when capture proved impossible, and even to destroy discovered IEDs at a safe distance by sniping the device with their heavy .50BMG antimateriel rifles.

IEDs would also be a physical hindrance to the sniper teams as they attempted to infiltrate into an area. The insurgents would place IEDs at so-called "vulnerable points," such as fords across streams or in compound alleyways they knew Coalition forces would likely cross. The Taliban also employed IED belts in defense of their base areas. These IED belts are, as the name suggests, concentric rings of IEDs in the manner of a minefield. They cannot be safely crossed without specialist Combat Engineer or EOD (Explosive Ordnance Disposal) support.

Snipers in Afghanistan were also faced with many of the same unique challenges every infantryman faced in that inhospitable and war-torn country. First and foremost, their shooting was governed by a set of rules that dictated when and at what the snipers could shoot. This differed between nations and sometimes between operations. For instance, it is generally regarded that U.S. snipers were given more flexible ROEs than other Coalition nations as they operated directly under U.S. command under OEF (Operation *Enduring Freedom*), rather than under NATO command as part of ISAF (although, confusingly, some U.S. troops were also under ISAF command).

Exceptions or temporary changes to the ROE could be applied for, as when, for example, the British conducted a planned offensive operation such as to capture an insurgent base area (areas where the insurgents stored their weapons, ammunition and explosives and that served as a sort of local headquarters for the insurgents in the region). In this case, they would be granted more lenient ROEs that allowed lethal force to be employed against insurgent spotters, known as "Dickers" to the British, who might not be armed but would be a using a walkie-talkie radio or cell phone to correct fire from insurgent mortars, marksmen or rockets, detonating an IED or informing on the route and direction of a patrol.

Such offensive ROEs also allowed for preparatory airstrikes or artillery barrages that were normally highly restricted. The basic ROE followed by Coalition snipers, however, meant they could only engage an individual who had displayed what was termed "hostile intent." The definition of hostile intent was of course something of a movable feast.

Typically snipers would also be required to PID or positively identify their targets as insurgents displaying hostile intent. This normally meant waiting until a weapon could be seen on a suspect individual before they could fire. Certain ROEs also required a minimum of two observers of a target before the target could be engaged. Despite these restrictions, snipers were perhaps the ultimate answer to collateral damage considerations in Afghanistan. They could surgically engage and kill enemy combatants with virtually zero risk of inflicting inadvertent casualties on the civilian population or infrastructure. They were also remarkably cost-effective when compared to other alternatives like the Hellfire missile or JDAM (Joint Direct Action Munition) guided bomb. "Snipers do not damage buildings or threaten the security of the indigenous population. They are a fantastic asset," explained one British officer who served in Helmand.[3]

In the SOF world, snipers operated under yet another set of ROEs depending under which command a specific operation fell. A retired

U.S. special operations officer told the author that he would invite his operators to briefing sessions with military lawyers when any such changes in ROEs were put into effect. The soldiers were encouraged to candidly ask about specific scenarios when they could or could not fire, allowing the operators to become empowered to make the call and not being forced to rely on command authorization when they were out on the ground. Any such delays could prove fatal for the SOF sniper.

The environment in Afghanistan would also pose its own set of problems for Coalition snipers. The terrain ranged from imposing snow-capped mountains in the north to the Green Zones and arid deserts of the south. Snipers had to be prepared to operate in all manner of terrain, much of it to the enemy's advantage. This made infiltration and avoiding compromise especially difficult. The fact that the insurgency used local sympathizers as look-outs only added to the danger.

An Estonian sniper in Helmand interviewed by veteran journalist Carl Schulze succinctly explained the problem:

Due to the thick vegetation and countless irrigation channels that provide good cover there is always a risk that a local farmer will emerge from somewhere and bump into us. They know their land well and notice the smallest changes. The problem is that you never know on which side they are so if we are discovered the mission fails. The Taliban are not stupid, they know that we can do almost nothing when they are not armed, therefore they conduct their reconnaissance camouflaged as unarmed farmers.[4]

An Australian sniper was blunter in his estimation of the challenges:

We stood out like dog's balls. Any strangers were instantly identified, which made working out of hides or in overwatch positions around the Green Zone bloody tricky, and hiding in the dasht (desert) in daylight often impossible.[5]

A Danish sniper agreed:

Even in the dense vegetation of the Green Zone, it is quite difficult to remain undetected. Farmers' children often emerge literally from nowhere. That makes stalking extremely difficult and dangerous. Once the children know where you are it will not take long for the Taliban to know as well.[6]

However, snipers were an effective way of beginning to understand the dynamics of the local situation.

An Australian officer quoted in *One Shot Kills: A History of Australian Army Sniping* commented on the value of his snipers in this regard in 2010: "my snipers were an important 'force multiplier'. It was not just their marksmanship that I relied upon; it was also their value as an intelligence asset and their ability to detect changes in the complex environment in which we operated."[7]

Environmental factors also influenced the selection of the weapons and ammunition Coalition snipers would deploy. The Russians discovered during the Soviet Afghan War that longer-range engagements were not at all uncommon. Their snipers in Afghanistan routinely zeroed their PSO-1 optics for 300 or 400 meters (329 or 437 yards), although shots out to 1,000 meters (1,094 yards) were recorded. In the Green Zones of irrigated land alongside major water courses, the ranges were reduced sometimes to less than 100 meters (109 yards) due to the dense nature of the foliage and trees and the civilian infrastructure in the area. Many civilians who earned their living through farming would live in compounds in or near to the Green Zone. Many of these structures would be used by the insurgency as firing points.

Many snipers serving in Afghanistan found that the lighter 5.56x45mm weapons carried by their security teams and fellow infantrymen were not providing enough penetrative punch when battling insurgents in the Green Zones or among the mud brick *qualas* that made up Afghan compounds (a *quala* is an Afghan adobe building). The 5.56x45mm would simply not reliably penetrate through the dense foliage or *quala* walls (in fairness to the 5.56x45mm, many of these walls would shrug off the high-explosive dual-purpose 30mm cannon round from the Apache gunship). Indeed, 5.56x45mm caliber weapons were also found not to be effective at the extended ranges the Taliban often favored.

The Taliban were cunning opponents and tried whenever possible to ambush Coalition forces at either very close range, as it negated close-air support and mortars, or at distances beyond roughly 500 meters (547 yards), to reduce the effectiveness of the smaller-caliber weapons carried by most infantry. The standard U.S. issue M4 carbine has an effective combat range of between 300 and 400 meters (329 and 437 yards) at best. Additionally, due to the short barrel of the carbine design, the 5.56x45mm round rapidly loses velocity beyond around 200 meters (219 yards), although recent innovations in bullet design have improved this.

The standard squad automatic weapon or light machine gun carried by Coalition forces, the Minimi Para, suffered a similar problem. One British officer with Afghan experience argued the Minimi was more of a morale boost than a particularly potent weapon beyond a couple of hundred meters. The insurgents of course understood these limitations and took advantage of them whenever possible.

A French sniper told the late Greg Roberts:

Here there are two types of combat. The first represents 75% of our TIC (Troops in Contact) and take part in the "green." The rest happen in the rocks, where our firepower gives us an advantage ... but never underestimate the enemy. The Taliban watch the Internet in their training camps in Pakistan and they have at their disposal instruction manuals and booklets so have a perfect knowledge of our material and equipment. They have also started to use a sniper team with spotters, armed with the Dragunov, but now we have received the HK417 technically we have the better equipment, though the enemy is cunning and ready to do anything to achieve his goals.[8]

As we will discuss later in this chapter, the concept of the Designated Marksman and the semiautomatic DMR (Designated Marksman Rifle) and sniper rifle both came into their own in Afghanistan thanks to the unique challenges of the environment.

Many nations contributed to ISAF. Among those involved in some of the heaviest fighting of the war was the Danish contingent. The Danes operated snipers alongside the British in Helmand. Their snipers were organized at company level with two two-man sniper teams available for tasking to individual infantry platoons or to operate on sniper-initiated ambushes and similar missions against IED emplacers. Danish Special Forces Jaeger Corps snipers also operated in Helmand. For instance, the initial Danish commitment in 2006 of a mounted light reconnaissance unit was soon reinforced by eight Special Forces snipers when the Danes were defending the town of Musa Qala, as the Danish conventional forces initially lacked a trained sniper capability.

"One of our main tasks is to provide overwatch and security for patrols. We are able to engage, with high accuracy, targets that are well outside the range of the assault rifles of the patrol members," explained one Danish sniper to Carl Schulze in 2011. The Danes provided patrol overwatch, including for cordon and search operations, along with supporting Coalition EOD teams to ensure they were not targeted by insurgents as they carried out their vital work. They were also involved in training the ANA along with the British and conducted extensive and far-ranging reconnaissance missions across the province. In many cases on these operations they used their surveillance cameras rather than their rifles ("we used cameras to take pictures of buildings that would later become subjects of missions"), although they were also employed in the war against the IEDs ("[we] were employed in areas where IEDs were planted regularly in order to hunt down the guys planting them").[9]

The Danes carried the excellent bolt-action .338 Lapua SAKO TRG42 with Schmidt & Bender optics (known as the Finskyttevaben M/04 within the Danish Army). Their sniper sections were issued the 7.62x51mm

HK417 with 20-inch (508mm) barrels during 2009 to 2010 to supplement their bolt-action platform. They discovered that with the reduced ranges encountered in the Green Zone, and both the immediacy of targets appearing and also the fleeting nature of those targets as they dashed into the thick foliage, another approach was required. The semiautomatic HK417 equipped with a Brugger & Thom suppressor gave them the ability to deliver fast follow-up shots and engage numbers of enemy at close range.

As other armies also were discovering, the semiautomatic sniper rifle could also be used as an assault rifle when the sniper was forced to operate as an infantryman, as for example during compound or *quala* clearances. The Danes also discovered that deploying the traditional two-man sniper team in Afghanistan was inviting trouble. Instead they deployed with a minimum of a pair of two-man sniper teams, often with a complement of infantrymen as security.

British snipers in Afghanistan were typically deployed either as part of a Fire Support Group (FSG) for a specific mission or attached in pairs to platoon- or company-level patrols. In the latter case they would tend to move with the TAC HQ or tactical headquarters element of the patrol until dispatched to a particular area of the battlefield by the commander of the operation, who would have tactical control of the snipers. When attached to the TAC HQ the snipers could also provide force protection for the command element, freeing up another infantry fire team to maneuver against the enemy. Such a grouping also made sense as it let the attached forward observers and JTACs enjoy the benefits of the enhanced optics of the snipers.

During one such operation during the Royal Anglian Regiment's 2007 tour of Helmand, the lead platoon managed to surprise a half-dozen insurgents who had let their guard down and were milling about unaware of the British presence. The snipers were ordered to simultaneously engage the insurgents as the mortar fire controller sitting next to them called in a barrage of 81mm mortar bombs on the area. After using the laser rangefinder to ascertain the range at 600 meters, the snipers prepared to take their shots.

Colonel Richard Kemp gave this detailed account in his excellent book *Attack State Red*:

[Lance Corporal Tom] Mann brought the crosshairs on to the center of the man's shoulders. If he had been firing the L96, he would have gone for a headshot, to guarantee a kill. With the much more powerful .338 Lapua Magnum, that would also be virtually guaranteed if he got him anywhere in the torso, and going for the shoulders gave a better chance of hitting. If his aim was slightly high he would get him in the head, if low he would hit in the chest.

He squeezed with his whole hand and the rifle jerked back into his shoulder. Three quarters of a second later the .338 bullet ripped open the upper body of the man with the AK47; he was flung back and crumpled to the ground in a heap.[10]

The other sniper narrowly missed with his first shot, but hit his target with his second. Within moments, six insurgents lay dead as the mortar bombs began to rain down to ensure the complete destruction of the insurgent unit.

The advantages of coordinated simultaneous shooting followed by mortar strikes were further spelled out by the British sniper, Monty B, who confirms it was a technique British snipers were trained in:

with all (sniper) pairs simultaneously engaging their respective targets at the same time, on command from the platoon commander on the ground, the threat will be neutralized instantly. Hopefully all targets will drop at the same time but it may be necessary to finish off the area with a fire mission e.g. by dropping mortars or artillery onto the position. Also destroying any kit and equipment including ammunition, food, soft-skinned vehicles, light armor or crew-served weapons will remove them from the enemy and help to deny him the future use of that position.[11]

Often British snipers would be pushed out to guard the flanks of an advancing patrol, as they could often spot encroaching Taliban long before the average soldier with his SUSAT (Sight Unit Small Arms, Trilux) or even ACOG (Advance Combat Optical Gunsight)-equipped L85A2. The SUSAT and its later ACOG replacement offered only a fixed 4-power magnification. They would also provide sniper overwatch on such operations, covering the infantry sections as they bounded forward, although this role was increasingly served by the Platoon Marksman, the British equivalent of the Designated Marksman. Once the forward sections reached the limit of sniper or marksman overwatch, they were halted while the snipers moved position to again cover the patrolling troops.

As mentioned, British snipers would also often deploy as part of a dedicated Fire Support Group or FSG. An FSG might contain a mix of vehicles such as WMIK Land Rovers (Weapons Mount Installation Kit, an armed Land Rover) or M-WMIK Jackals (a 4x4 off-road patrol vehicle) mounting heavy weapons platforms like the .50BMG M2 heavy machine gun or the 40mm grenade machine gun, and crew-served weapons systems from the Support Company. These normally included Javelin antitank guided missiles, a 51mm or later 60mm infantry mortar, 7.62x51mm L7A2 General Purpose Machine Guns (GPMGs) and sniper teams.

The general concept was for an FSG to move into a previously identified and reconnoitered position that provided good overwatch opportunities to cover a planned patrol or cordon and search mission through the Green Zone, for instance where lateral visibility was often poor. The FSG would be able to see targets that the infantry platoon or its multiples (typically two infantry sections) simply could not see.

Coupled with a tactical UAV such as the Desert Hawk, the system worked very well. The FSG could both positively identify its own targets and engage them (once they had ensured deconfliction had occurred with the infantry on the ground to avoid the possibility of a blue-on-blue fratricide), or it could fire on targets as directed or marked by the infantry unit on the ground. With the mix of weapons platforms available, the FSG commander could choose the right weapon for the job.

Not surprisingly, the snipers were often tasked, as they could deliver precision fire with near zero risk of collateral damage. The sniper teams would also act as spotters for the Javelin teams using their issue 40-power spotting scope. Both the Javelins and snipers were regularly used to pin an insurgent element in a particular area such as an identified enemy *quala*, stopping them from withdrawing so they could be destroyed by mortar, artillery or airstrikes.

In a contact, British snipers, in concert with all Coalition snipers, would prioritize the typical sniper targets: obvious leaders, heavy weapons and enemy marksmen. With their 7.62x51mm L96A1s, and more so with the .338 Lapua L115A3s that were adopted in 2007, they could actively target the longer-range Taliban weapons systems – like the 7.62x54mm PKM light machine gun – that were beyond the reach of the standard infantry small arms. The L96A1 was sometimes still preferred in the Green Zone, as the engagement distances were so reduced and the weapon was certainly much lighter than the L115A3. The .338 Lapua was preferred when engaging targets behind cover or at extended ranges. The bolt-action .50BMG AW50 was also available and would be occasionally employed, although typically only on operations where there was a high probability of insurgents in heavy cover such as in bunkers and trench lines.

The Dutch Army and Marines were in fact one of the earliest to procure .338 Lapua rifles for their snipers. In Afghanistan, they used the .338 Lapua Accuracy International Arctic Warfare Magnum to great effect. Like the Danes, the Dutch also adopted the 7.62x51mm HK417 semiautomatic as a supplementary weapon to the bolt-action .338. Until the arrival of the semiautomatic L129A1 marksman rifle in 2009, however, British snipers lacked a weapon capable of quickly filling the air with lead.

Their snipers instead developed a technique known as *rapid bolt* that was unconventional but worked. The sniper would place his rifle in his left hand and rapidly work the bolt with his right, manipulating the trigger with his left trigger finger. It enables the sniper using a bolt-action platform to rapidly fire a volley of shots and is often used when employing the *trap* or *ambush* method of leading a target.

British commanders who understood the unique skills of the snipers would take advantage of their reconnaissance and surveillance training to get "eyes on" a potential target from a covert OP. They would also be employed with a dedicated signaler (radio operator) in an overwatch position that could feed information directly to the senior commander. When used alongside tactical UAV systems, a well-placed sniper team could give their leaders a virtual three-dimensional image of the battlespace. British snipers would also sometimes be deployed alongside Light Electronic Warfare Teams or LEWTs who could monitor insurgent communications and triangulate their locations. Having the snipers nearby shortened the communications chain between sensor and shooter.

During the Welsh Guards' 2009 tour of Helmand, a sniper pair from 4 Rifles attached to the Welsh Guards Battle Group, Serjeant Tom Potter and Rifleman Mark Osmond, forward deployed to Patrol Base Shamal Storrai, a newly constructed PB. They were both members of the team that had taken the British Army Sniper Championship in 2006 and had accumulated 30 kills between them during multiple tours in Basra, southern Iraq.

In some 40 days at the PB, the pair racked up 75 kills, with most shots taken from a static hide within a *sangar* near the front gate. The pair used sound-suppressed 7.62x51mm L96A1s, although the rifle had been officially replaced by the .338 Lapua L115A3 some two years previously. According to author Toby Harnden, the majority of the shots were made at extended ranges of up to 1,200 meters, although one kill was achieved at 1,430 meters, which is doubly impressive considering the limitations of the 7.62x51mm round. Harnden reports that the pair, knowing the local weather conditions, went for headshots up to 900 meters and the center of body mass beyond that.

One of their most celebrated shots, as previously mentioned, occurred on September 12, 2009, when a known insurgent leader was spotted riding pillion on a motorcycle with another insurgent. As a British patrol was nearby and the rules of engagement at the time considered the walkie-talkie carried by the Taliban leader as a "hostile act," one of the snipers engaged the target at a little under 200 meters. The round struck the insurgent commander in the head, killing him instantly, but then passed through his skull, also hitting the second insurgent in the head and killing him.

Their techniques were born of necessity: often needing both rifles in action due to the numbers of targets, the snipers spotted for each other using the rifle's optics rather than a traditional spotting scope. They also used a variation of the *trap* or *ambush* technique, with one sniper announcing a moving target that he had spotted along with probable direction or destination, allowing the second sniper to set up the shot in advance.

The pair also tended to dispense with the dedicated spotter role unless the target was at extended range, such as over 1,200 meters, when one of the pair would man the spotting scope to call in adjustments. They also worked closely with an Afghan interpreter, who used an ICom scanner to covertly monitor insurgent radio communications. He would feed intelligence to the snipers, allowing them to second-guess the insurgents and set up shots.

The effect on the local population was surprising. The death toll from the snipers forced the Taliban from the local village as it was too dangerous to stay. The closest insurgent firing position was pushed back some 1,200 meters thanks to the snipers. The ultimate reward was the arrival of local civilians at the patrol base, happy to talk with and engage with the British forces, as they no longer felt the constant intimidation of the Taliban. This effect was even more pronounced when the heavier-caliber rifles such as the .338 Lapua were employed, as the insurgents soon learned to stay well out of range. Typically this meant retreating to beyond 1,500 meters.

During the same tour, another sniper managed five kills during a single contact, with each of the kills at distances over 1,600 meters. The sniper, Corporal Chris Fitzgerald, was deployed in a permanent hide site in a *sanger* at a patrol base providing sniper overwatch for an infantry patrol. He was accompanied by his spotter and two U.S. Marine forward observers who were passing on intelligence from a live video stream they were watching from a UAV overhead (such receivers, known as Rover terminals, allow soldiers on the ground to see what an aircraft or UAV sees). Soon a Taliban RPG team was spotted, about to engage the patrol.

Harnden quotes the sniper: "We didn't really have data beyond 1,000 meters so it was pretty much working off the cuff. The suppressor was bringing up the elevation. I had a 19 MPH wind and I was shooting over compounds, which also affected the wind. There was a lot of luck in it."[12] The sniper took nine rounds to hit the RPG operator with a round to the chest at an astounding 1,764 meters, guided by his spotter due to the heat haze. He followed this by dropping an insurgent machine gunner carrying an RPK, two more with AK47s and an insurgent leader using a cell (mobile) phone. Between the three snipers, they notched up 131 kills, three quarters of the total enemy deaths inflicted by British forces during the tour.

Another British sniper, a Royal Marine Commando from the Brigade Patrol Troop, allegedly holds the record for the most sniper kills, with 173. The majority of these were apparently chalked up during the Marines' 2006–2007 *Herrick V* tour while he was a member of the Brigade Reconnaissance Force (BRF). The BRF acted as the eyes and ears for Task Force Helmand, conducting both long-range mounted reconnaissance and shaping operations designed to confirm insurgent safe areas and test their responses to incursions into their territory.

In terms of longest-range engagements in Afghanistan, there had been more than a few contenders. The non-classified record is still held by

Sergeant (then Corporal of Horse) Craig Harrison of the Household Cavalry Regiment in 2009 at Musa Qala. Harrison engaged an insurgent PKM team at 2,475 meters with his .338 Lapua Magnum L115A3. Due to the incredible range, it took the sniper nine rounds in total to range in on and kill the pair of gunmen with two center-of-body-mass hits. Sergeant Harrison's highly recommended autobiography, *The Longest Kill*, also recounts the tale of a Taliban Dicker that he engaged from almost 3,000 meters, firing a total of 13 rounds to keep the spotter suppressed. The harassment fire apparently worked, as Taliban communications were intercepted lambasting the poor spotter for not providing information. The spotter retorted that he could not as every time he stood up someone shot at him!

That particular engagement, which led to an ambush of a Jackal mounted patrol and Harrison's world record shot, is illustrative of the ranges the sniper would encounter in Afghanistan. After Harrison engaged the spotter at some 3,000 meters, he killed two Taliban at 760 meters followed by an RPG team at 700 meters before engaging another RPG team at 1,000 meters. Finally, he famously engaged the PKM team at 2,475 meters. The reader may note the use of the term "non-classified record." As we will see later, the classified record is held by an Australian Commando sniper team that made an incredible 2,815-meter shot in Helmand in 2012.

One of the other most discussed sniper shoots of the Afghan campaign occurred on December 14, 2013, again in Helmand. A British sniper, a lance corporal in the Coldstream Guards, was about to make history. He had already killed an insurgent PKM gunner at a very respectable 1,340 meters (1,465 yards) with his L115A3 earlier in the tour. The sniper and his spotter had identified a group of insurgents, including suspected suicide bombers, from the commanding heights of the Sterga 2 Observation Post, a permanent OP that overlooked both Highway One and Route 601, two main thoroughfares that were constantly under the threat of IEDs. The insurgents also often attacked the ANP checkpoints on the roads.

With the insurgents spotted, a dismounted patrol of ANA and British infantry was dispatched to attempt to flank the insurgents. They were contacted by the enemy and a firefight erupted. Some 850 meters (930 yards) away, the sniper watched as an insurgent, carrying an RPK light machine gun, emerged from a creek bed and moved into view. His commander explained what happened next: "They were in contact and he was moving to a firing position. The sniper engaged him and the guy exploded. There was a pause on the radio and the sniper said, 'I think I've just shot a suicide bomber'. The rest of them were killed in the blast."[13]

In fact, the sniper had killed six insurgents with a single .338 Lapua Magnum round. The vest on the lead insurgent had detonated and exploded, killing the unseen insurgents moving up behind him. After this, the insurgents abruptly broke contact and retreated away. As the infantry patrol cleared through the site, a second suicide bomb vest was discovered near the remains of the six insurgents. The patrol speculated that the enemy force included Pakistani terrorists intending to blow themselves up at a nearby ANP checkpoint.

The U.S. Marines also deployed their Scout Sniper teams to Helmand when they rotated into the province in 2007 before finally taking responsibility for large tracts of Helmand in 2009. The Marine sniper teams would typically deploy in eight-man elements, both as a reaction to their experiences in Iraq where a number of Marine snipers were ambushed (as we shall see in the following chapter) and as a protective measure against Taliban swarm tactics. As the name implies, this was a tactic used by the insurgents to try to isolate and overrun a Coalition unit. The insurgents often attempted to separate fire teams (typically of four men) so they could attempt to swarm them. The technique had the added benefit from the insurgents' point of view of virtually negating close-air support, as the enemy was often too close to the "friendlies" (a term used to describe any Coalition unit) for pilots to be comfortable dropping ordnance.

Marine Sergeant Ben McCullar, a sniper with 3rd Battalion, 2nd Marines, noted that by 2011, the Taliban in Helmand, and particularly where he operated around the oft-contested Musa Qala, had learned the concept of weapon overmatch and stayed out of the effective range of the majority of the Marine sniper teams' weapons. "They'd set up at the max range of their PKMs and start firing at us," he explained. "We'd take it until we could call in [close air support] or artillery."[14] Despite these difficulties, his 33-man Scout Sniper platoon killed a confirmed 185 insurgents by the completion of their seven-month tour in Helmand.

The Americans also found themselves operating in the east along the border with Pakistan. The region provided its own set of unique difficulties for the sniper. First, as can be seen in the award-winning documentaries *Restrepo* and *Korengal*, the imposing terrain forced snipers to brush up on their high- and low-angle shooting. As we explained in the previous chapter, shooting up or down mountains has a drastic effect upon the bullet and the set of calculations needed to ensure accuracy. It also offers a tremendous advantage to the insurgent, as the natural camouflage of the region makes it incredibly difficult to spot a target beyond anything but close range. Even in the open, the insurgent is easily hidden against the camouflaging tones of the rocky terrain.

The insurgents would often ambush any patrol that attempted to advance through the inhospitable mountains and valleys of the Korengal. The late Army sniper Sergeant Russell M. Durgin was a casualty of one such ambush. Durgin was leading his six-man sniper element when they were contacted by a numerically superior enemy force. Durgin was killed while covering the withdrawal of his small team. In another similar ambush, two insurgents tried to drag away an unconscious and wounded American soldier before they were shot and killed.

The U.S. Army established temporary units known as "kill teams" that combined a Combat Observation and Lasing Team (or COLT) alongside a sniper element. A COLT typically comprises three soldiers equipped with a range of optics including thermal imagers along with laser designators and radios to call for fires or close-air support. The sniper element was normally a full sniper section of three teams of three equipped with the 7.62x51mm M24 and the .50BMG M107. The concept was for the snipers to engage identified targets within range that were either self-spotted or spotted by the COLT. For larger concentrations of enemy or for targets beyond the capabilities of the M107, the COLT could call in mortars, artillery or air strikes.

One such kill team was employed to support an offensive operation called Operation *Mountain Thrust* in the mountainous northeast of Afghanistan in June 2006. The objective for the kill team was to observe a number of areas known to be frequented by insurgents and engage any identified enemy who were forced from the valley by Operation *Mountain Thrust*. The team infiltrated after a difficult climb to their previously identified observation posts. Upon reaching the heavily wooded area, they also discovered a well-trodden trail going right past the location of their OP. Based on the fact that they would only be in position for a short time, it was decided to go ahead and establish the OP, but also to place security covering the trail.

Soon they discovered that they would in fact be in place for several days, as the start time for *Mountain Thrust* was delayed. As they had packed light in terms of provisions, an aerial resupply was requested. Unfortunately the resupply was dropped from a helicopter hovering directly over their position. One soldier soon spotted an Afghan now watching them through binoculars. Knowing they'd been compromised, the kill team nonetheless decided to stay put, as they felt any movement to extricate themselves would end in ambush.

In a co-ordinated attack, an estimated 50 insurgents opened fire from three directions on the kill team just before dusk. Two soldiers were almost immediately killed. Staff Sergeant Jared Monti of the COLT made a desperate call for supporting indirect fire, and was killed trying to rescue a wounded soldier caught in the open. Monti attempted to reach him three times, but was killed by an RPG on his last attempt. Monti's actions resulted in a B1B bomber arriving on station to bomb the firing positions the insurgents had established.

The sniper section engaged and killed insurgents as they showed themselves, sometimes only by their muzzle flash. Artillery and mortars were called in by the COLT that eventually helped stop the insurgent assault, dropping mortar bombs and artillery shells on the ridges overlooking the OP. Although the soldiers had killed 26 insurgents, four Americans had died, two tragically in a helicopter accident as the wounded were extracted via a rescue hoist that snapped. Staff Sergeant Monti was posthumously awarded the Medal of Honor for his actions.

Another American sniper distinguished himself in the rugged terrain of eastern Afghanistan. U.S. Army Sergeant Nicholas Ranstad wrote himself into the record books with a 2,288-yard (2,092-meter) kill with his .50BMG M107 in January 2008. Ranstad had been conducting shooting practice from a combat outpost against some natural features in a nearby ravine. For each shot he added the details including the wind, distance and the angle of the shot to his range card. Now four armed insurgents had appeared in the ravine.

One stepped out into the open and Ranstad fired. The .50BMG missed, impacting at the insurgent's feet. The man made it to cover but one of his compatriots appeared, evidently trying to locate the origin of the shot. Ranstad fired a second round, this time impacting the center of mass and dropping the insurgent, registering the longest-range kill by a U.S. sniper in Afghanistan to date.

The French found themselves operating in similar terrain and, like the Americans, fell back on their .50BMG antimateriel rifles too. Their snipers were organized into light and heavy teams. The light teams carried the 7.62x51mm FR-F2 or HK417, while the heavy teams used the .50BMG PGM Hecate II. In Afghanistan the French deployed their sniper elements together, in some cases with additional firepower added as they often operated in a reconnaissance role far from other friendly units. As an example, a pair of light and heavy sniper teams would operate together along with two spotters who, armed with the 5.56x45mm Minimi light machine gun, provided additional security.

The Estonian sniper contingent in Afghanistan who served alongside the British in Helmand also used the French .50BMG PGN Hecate II mounted with a Hensoldt 6 to 24 mildot variable optic. The Hecate was often seen operationally fitted with a surprisingly effective sound suppressor that could reduce the weapon's report by as much as 25 decibels, certainly not silencing it but making it less of a chore to shoot for the sniper and perhaps disguising the source of the fire from the target. For targets out to around 1,000 meters (although Estonian snipers in Helmand made kills out to 1,300 meters with the platform), they employed the .338 Lapua SAKO TRG-42 fitted with the same scope and also with a detachable suppressor (which had replaced the French MAS FR–F2 in Estonian service). Their spotters surprisingly carried the 7.62x51mm American M14 with Schmidt & Bender 3 to 12 variable power optics, which they used to engage insurgents out to 700 meters.

More recently, a U.S. Army sniper team armed with the newly adopted .300 Winchester Magnum M2010 was part of an infantry platoon providing overwatch and security for a Shura (a meeting between village elders and representatives of ISAF or the Afghan government), again in mountainous eastern Afghanistan. As they moved into position, the platoon was engaged by enemy small-arms fire and 82mm mortar rounds. The spotter identified an insurgent manning a PKM at a distance of between 800 and 900 meters, well outside the engagement range of the 5.56x45mm M4 and pushing the capabilities of even the 7.62x51mm M24.

With his second shot, guided by adjustments called out by the spotter, the sniper shot and killed the PKM gunner. Moments later, the spotter positively identified a number of insurgents attempting to flank the platoon's position. As the spotter explained: "At that point I suppressed the enemy's movement with my weapon as (the sniper) oriented himself to the enemy. He held 4 and was a first round hit eliminating the threat. After that the enemy pulled back completely."

As we shall see as we examine the specific types of sniper operations conducted in Afghanistan, both the heavy caliber and the semiautomatic rifle became predominant. We will also see that the core skills of the sniper, his fieldcraft and his single-shot marksmanship, were still as important in Afghanistan despite the counterinsurgency nature of the conflict. In fact, the war in Afghanistan saw snipers employed upon operations that were not considered within the sniper's domain until very recently, causing a drastic rethink of sniper tactics.

SNIPER OVERWATCH MISSIONS

The sniper overwatch mission was perhaps the most common of the sniper operations in Afghanistan. *Overwatch* is simply to maintain a position of vigilance over an area or over another Coalition unit. Snipers would deploy on sniper overwatch to provide both early warning of insurgent intentions and also precision supporting fires once those enemy intentions were confirmed. Sniper overwatch could be conducted from a static hide site, either in the field or from a HESCO bunker at a Coalition PB or FOB. It could also be provided from the air or alongside an infantry patrol with the snipers covering their next move or bound.

In Afghanistan snipers would provide overwatch for Shuras, or for what is known in Army speak as a KLE or Key Leadership Engagement. They would also commonly provide overwatch of Afghan National Police checkpoints and outstations that were the frequent target of the insurgents. One fairly unique sniper mission in Afghanistan was the covert overwatch of local markets. Markets, the central hub of Afghan towns, would often be the targets of Taliban attacks, particularly later in the war as they increasingly employed Pakistani suicide bombers in an effort to destabilize any support for the central government and security forces. ISAF sniper teams would often be tasked with establishing a covert hide to watch for insurgents laying IEDs in the markets or for the approach of suicide bombers.

Most sniper overwatch missions in Afghanistan were not strictly kinetic: they were primarily for intelligence gathering. "Our snipers' secondary task is to gain information on the enemy. They are highly trained in the gathering of information, observation and deduction and the study of the pattern of life in the target areas," said Australian sniper Nathan Vinson.

When snipers deployed into a new area they would request data on all known previous firing points used by the insurgents and immediately range these targets with their laser rangefinders, as they soon learned that the Afghans would often reuse firing positions. The ranges to these points would be written up on range cards either carried by the spotter or taped to the stock of the sniper's rifle for easy reference. After this they would begin to observe and understand the local area.

British sniper Monty B explained in his autobiographical *A Sniper's Conflict*:

The many hours spent up on the roof conducting observation and data-collecting were never wasted: we watched the pattern of life, the daily routine of the civilian population around us, figuring out who lived in which compounds, which local farmers worked in which fields and so on. Then there were the locals from outside our area.

We meticulously observed their routines and habits throughout the day as they moved in and out, also paying attention to what they were wearing and with whom they were associating. Just as important was to gain information on local use of motorbikes [a key Taliban transportation method] and other road vehicles as they transited through our area. What were they transporting? And to what destination?[15]

Often covert hide sites were used to lessen the chance of compromise by a curious local or goat and to reduce the chances of spooking the insurgents with an overt presence. A common tactic, however, would be for Coalition snipers to deploy on an overt patrol with an infantry platoon but to quietly split off and establish a covert hide site, staying behind to monitor and surveil an area of interest. Usually (in a reflection of tactics first developed in Northern Ireland and refined in Iraq) a quick reaction force of infantry was stationed nearby, ready to come to the rescue should the sniper team be compromised.

Such an incident occurred on Operation *Red Wings* in 2005. Now famously known from the book and film *Lone Survivor*, *Red Wings* was a U.S. Marine operation to disrupt insurgent activities in the notorious Pech District of Kunar Province before the then upcoming Afghan elections. As part of the operation, a four-man SEAL reconnaissance element was infiltrated into the area to identify an insurgent leader. The team was soon compromised by civilian goat herders, who informed the insurgents of their presence. They were ambushed by a numerically superior force and three of the SEALs were killed. One, Navy Hospital Corpsman Second Class Marcus Luttrell, the team's medic, survived and was shielded by Afghan civilians until he was rescued by U.S. Army Rangers.

The book and film of the incident make much of the question of whether the SEALs should have killed the goat herders. The book in particular also criticizes the rules of engagement, arguing that because the SEALs would likely face criminal charges if they had killed the goat herders, the ROEs in effect damned the team to failure. Critics have argued that perhaps the SEALs could have simply restrained the civilians, securing them to a tree and gagging them before they attempted to withdraw to their emergency rendezvous point. Similar challenges have faced other sniper teams in the past, including a Green Beret team during Operation *Desert Storm* that also allowed the civilians to go free. They also immediately notified Iraqi forces.

Snipers are also employed overtly to provide overwatch. British snipers were deployed from the first Helmand tour in 2006 as part of the fortified defenses of the platoon houses that were located in some of the province's worst trouble spots. As daily patrols would often not be able to venture further than line of sight to the platoon house before running into the enemy, the snipers could provide precision fire support as patrols broke contact. Colonel Stuart Tootal, commanding officer of 3 Para during the break-in battles of 2006, described how his snipers were employed:

If artillery, bombs and Claymores (mines), were the blunter instruments used in the defence of Sangin, the employment of snipers provided a more surgical tool. Each of the company groups had a sniper section attached to it consisting of six men from the battalion's Sniper Platoon ... The snipers stationed themselves in positions that gave them a commanding view of the ground around the district center and worked in pairs. The more experienced of the pair would act as the spotter.

One of their primary tasks was to counter the threat of insurgent gunmen attempting to conduct their own snipes ... In one particular incident, an intelligence report indicated that a Taliban marksman would attempt a shoot from a given location at a particular time. A sniper pair was tasked to cover the likely firing area and neutralize the threat when it materialized.[16]

The spotter noticed a figure slowly coming into view to line up a shot, over 1,000 meters away on the roof of a building on the outskirts of Sangin. The sniper fired a single round, the .338 Lapua bullet hitting the insurgent in the chest and killing him instantly.

Snipers also provided mobile sniper overwatch as part of platoon- and company-level patrols and sometimes for SOF operations. On one such operation, three U.S. Army National Guard snipers of the 1st Squadron, 221st Cavalry Regiment came close to being overrun by insurgents in September 2009 in Laghman Province. Deployed on a joint operation with a team from the U.S. Army's 10th Special Forces Group and an element from the French Foreign Legion, the Investigation and Liaison Detachment (an *ad hoc* team of snipers and reconnaissance specialists), their task was to establish covert hide sites to observe a possible weapons delivery from Pakistan under the guise of a local Shura. To maintain secrecy, the joint unit was inserted into an off-set landing zone by helicopter on a mountain ridge and patrolled for two hours through the rugged mountainous terrain until they established their observation posts on top of a rocky hill.

The Green Beret team split into three, manning a trio of observation posts, with the snipers and the French team in overwatch positions. The operation was uneventful until the team were preparing to exfiltrate. The landing zone had been ringed with Taliban insurgents who had quietly moved into position to ambush as the Coalition forces exfiltrated. What ensued was a fierce two-hour firefight that saw the small force attacked by more than 50 Taliban in human wave assaults. The American and French snipers and the Green Beret team ended up in two fighting positions desperately holding back the enemy while awaiting incoming air support.

One of the National Guard snipers raced through small-arms fire to assist the Special Forces who had called for sniper support. Incredibly, when he reached the Green Berets, he was told to head back as his weapon, a 7.62x51mm M14, wasn't equipped with an infrared laser (the Green Berets most likely wanted the sniper to identify and mark targets for them with the laser, visible only to those wearing night vision). Another sniper, this time with an infrared laser on his weapon, was dispatched to help the Green Berets fight off swarm attacks by the Taliban. To add to their worries, insurgent mortar rounds began to creep toward them.

One of the French snipers explained in an interview with the late Yves Debay:

There were fourteen others at my positon and the other position had nine men. Brutal fire erupted from everywhere from AK47s, PKM machine guns and RPG-7 rockets. Two U.S. Green Berets were immediately hit. The medic received a bullet through the calves and my best American friend was well shocked by a bullet in the hand, two bullets in his helmet and two in his flak jacket. The other side probably also suffered because we had six Minimi LMGs for 23 men.[17]

At one point, one of the sniper team spotted a large group of insurgents advancing up the hill they were positioned upon. The soldier fired off a whole drum from his M249 SAW, killing a number and sending the others running for cover, but still the insurgents kept coming. For a moment the U.S. snipers thought that they might not make it home:

"Man, I don't think we're walking away from this one," one reported thinking. Another said that he kept thinking of the consequences of capture: "My head's getting chopped off. I'm trying to think of how much ammo I had left ... That was my biggest fear, my head getting chopped off on the Internet."[18]

Low on ammunition, the situation was saved by the timely arrival of a flight of F-15E Strike Eagles who mowed down insurgents with multiple gun runs, followed by a pair of Apache helicopter gunships. The Apaches, using their thermal imagers, began to thin the Taliban ranks with their nose-mounted cannon and destroyed the mortar teams. Despite this, the tenacious insurgents would still step out from cover as the helicopters passed overhead and engage them with small-arms fire and RPGs.

Even when the French Caracal transport helicopters had landed and the soldiers were in the air and headed for home, they were fired upon by an RPG that airburst near the tail of the helicopter, peppering it with shrapnel. For a grim moment they thought they could add a helicopter crash to the horrors of the day, but the skills of the pilot kept them aloft and he nursed the helicopter, and its sniper team passengers, to safety. One of the escort French Tiger gunships spotted the RPG launch and proceeded to unleash 12 2.75-inch unguided rockets and some 400 25mm rounds into the area, suppressing any further insurgent activity.

Providing mobile sniper overwatch for conventional operations could prove just as dangerous. During the 2007 tour by the Royal Anglians, a battalion sniper engaged and killed at least seven insurgents during a single engagement, allowing an encircled patrol to extricate from an ambush east of the notorious Kajaki Dam without friendly casualties. A sniper with the battle group at the time, James Cartwright, commented on his first probable kill in his book *Sniper in Helmand* after the mounted patrol he was traveling in was ambushed:

Another RPG flew over the top of Strikey's WMIK bonnet just as he was shouting, "Over there! Reference that building with three windows, next to the tallest building!" I zapped it with my rangefinder, pinpointing the target at 628 meters. I took into account the wind aspect while thinking "this is it!" My heart was racing as I aimed and took up the first pressure on the trigger.

There was an incredibly bright flash as yet another RPG left that particular window and I immediately put two rounds in quick succession directly into it. I will never know if I killed the man with the RPG, but the firing ceased immediately. I guess he had either been hit and was now dead or I missed and he made good his escape.[19]

Australian snipers would provide similar mobile overwatch for their patrols in Uruzgan Province. In one battle in December 2010, a joint force of 15 ANA and 11 Australians from 5 RAR set out to conduct a presence patrol into the nearby village of Derapet. Among the Australians were two sniper pairs. One pair, equipped with the bolt-action .338 Lapua Blaser 93 Tactical 2, was sent up onto the high ground, while the other, carrying the semiautomatic 7.62x51mm SR-25, advanced with the patrol.

As the patrol entered the village, they noted a strong "combat indicator" – the civilians were packing up and leaving in a hurry. The overwatching snipers proved their worth by identifying a large number of armed insurgents headed toward the patrol. As they were taken under intense RPG and machine gun fire, the patrol broke contact under the supporting fire of the snipers, who held back the advancing insurgents long enough for a pair of Apache attack helicopters to be brought into action, breaking the back of the insurgent assault and allowing the joint patrol to withdraw without casualties.

One of the more unusual overwatch roles for Coalition snipers was to contribute to the safety of voters during the Afghan elections. Helmand, for instance, saw a tiny minority registered to vote, and even fewer turned up on the day thanks to Taliban intimidation. However, the British forces did as much as they could to ensure the security of those that did make the effort. A British sniper was employed to engage a

Taliban snap checkpoint that had been established on the approach road to a voting station.

Interestingly, the ten-man Taliban unit probably knew it was under surveillance, as they ensured that none of their weapons were carried openly and instead were hidden within the folds of their clothing. One insurgent did inadvertently expose his AK47 and the sniper took the opportunity, sending a .338 Lapua into the center of body mass and killing the man instantly. The other insurgents took to their heels and the road was cleared for local civilians to vote without Taliban intimidation.

Nathan Vinson recounts the Australian experience in Uruzgan Province: "In the early days of the deployments, the sniper teams were either deployed as overwatch of patrols or went out with the patrols on foot and integrated with the patrols. During the latter deployments, they were deployed much the same but with the additional task of training ANA marksmen as well."

The location of the overwatch tasking would influence the choice of sniper rifle carried.

If you were in an overwatch position then maybe the SR98 or .338 Blazer. But if you were down in the Green Zone you would leave the long rifle behind and take either the SR25 or the HK417. That would be the same for the Iraq style ops. It just depends on your task for that operation. The snipers of the Australian Army are very lucky nowadays, they have a suit of weapons to choose from – .50cal AW50, .338 Blazer, SR98 and the HK417.

COUNTER-IED MISSIONS

As we have noted, snipers played a pivotal role in the war against the IED in Afghanistan. Part of the sniper overwatch mission was to conduct overwatch of IED sites. Like firing points, the Afghan insurgents would often emplace their IEDs in the same locations, making it essential for Coalition forces to document the coordinates of any IEDs discovered.

These IED overwatch missions would often be carried out in conjunction with camera surveillance from observation balloons and mast-mounted sensors at Coalition FOBs. If an IED team was spotted by any of these devices laying an IED, they could then be engaged by the sniper team.

Snipers would also be used in sniper-initiated ambushes based on insurgent patterns of behavior. On occasion these would be based around elaborate deceptions to draw out the insurgents. During the Welsh Guards' 2009 tour, a deception plan was put into place with a faked breakdown of a Mastiff Protected Patrol Vehicle. A sniper element was covertly inserted to overwatch the area.

As the simulated recovery progressed, a three-man insurgent IED team arrived in the area and the snipers watched as they attached battery packs to devices, intending on triggering it as the patrol left. Author Toby Harnden described the incident in his groundbreaking book, *Dead Men Risen*: "The figures moved back and forth from the trees to the track, lying flat for up to 10 minutes as they armed IEDs. After reporting the movement back to [his commander], he [the sniper] was given authorization to fire. The next time one of the figures stood up, [the sniper] fired a single shot, hitting him in the chest and killing him instantly."[20]

Australian snipers were an essential part of the war against the IED in Uruzgan too. Colonel P. Connolly explained the role they played in 2009, giving a unique look at small-unit tactics of the period: "They were employed extensively on patrols within the green zone to 'satellite' the supported patrol and remain off-set from the main body to look for RC/CW IED 'trigger men.' Snipers were specifically employed for counter-IED operations, and to contribute a layer of intelligence, surveillance and reconnaissance for major deliberate operations."[21]

Snipers were also heavily involved in the campaign to target the bomb makers, their financiers and logisticians, along with their command elements, although this was primarily an SOF mission and as such will be covered later in this chapter.

COUNTERSNIPER MISSIONS

During large-scale offensive operations in Afghanistan, snipers were employed in their traditional role as hunters of enemy snipers. During one such operation in September 2009, Army snipers supported the ANA and SOF components from MARSOC (Marine Special Operations Command) as they attempted to clear the town of Shawan of insurgents. One sniper recounted the operation to Carl Schulze:

> One of my sniper teams took up a position on top of a roof from where they could cover the area in which we thought the enemy sniper might be hiding. When he opened fire again on a section of ANA we were able to identify his position. He was armed with a SVD Dragunov and shooting through a mousehole from inside a building.
>
> Just when our shooter was about to engage him, the team came under fire from somewhere else. Despite this, the shooter kept his weapon trained on the point where the target was hiding, then when the Taliban came up to take another shot – the range was 500 meters – [the sniper] quickly took aim and squeezed the trigger. The first round was short and punched into the wall below the target. Speedily he squeezed the trigger of his M110 SASS again and the second round hit the Taliban in the chest.[22]

A British sniper team was deployed in late 2006 in Now Zad to hunt an insurgent sniper that had been sniping at patrols leaving the District Center, the solitary outpost of the central government. What was thought to be a single shooter initially was actually four Taliban marksmen who soon lay dead at the hands of the British sniper.

Another British sniper team was dispatched to Patrol Base Silab in Nad Ali after a British soldier was shot in the head by a Taliban marksman. Thankfully the soldier was not killed, but the threat of another such shooting saw a sniper team deployed, which eventually spotted the insurgent lining up a shot on a British sentry. He was engaged and killed moments before he could squeeze the trigger.

COUNTER-IDF MISSIONS

One crucial role that snipers found themselves tasked with in Afghanistan (and later Iraq) was interdiction of enemy mortar and rockets. An American cavalry sniper described one such operation:

During our 2007/8 deployment our base, COP Wilderness, was regularly mortared with the Taliban all the time using the same firing position at the same time of day. [The sniper team leader] decided to ambush the mortar team with our two sniper teams. Around 09:30 hours on a Sunday morning the Taliban showed up, slightly later than usual.

[The sniper team leader] shot the rearward man of the three right between the shoulder blades with his M110. The distance was 715 meters. Shortly after, the other two members of the Taliban mortar team, who had taken cover between rocks, were killed by an airstrike conducted by AH-64 Apache attack helicopters that were guided onto their objective by the snipers via radio.

Snipers would be tasked to overwatch suspected base-plate areas awaiting insurgent mortar or rocket teams. As we shall see, in Iraq the base-plate area was normally distinguished by an actual base plate bedded into the ground, the theory being that the mortar team could arrive with the mortar tube itself, attach it to the base plate and bipod legs and fire a few rounds. They could then quickly break down the component parts and escape before Coalition attack helicopters could arrive or, if authorized, counterbattery by the Coalition's own mortars landed. In Afghanistan, the base-plate areas were normally simply the preferred firing location for the insurgent mortar teams, although in some areas foreign fighters went to the considerable trouble of cementing base plates into the ground.

Like IEDs and attacks by enemy marksmen, the coordinates of all indirect-fire attacks are important to have available to sniper teams because, as we've seen, the Afghans tended to reuse the same locations. Even more difficult to spot than base plates are the launch sites of the

Chinese 107mm unguided rockets the insurgents fire at Coalition PBs and FOBs. These are simply propped up on a pile of rocks or at best a jury-rigged bipod made from sticks and fired off in the general direction.

DESIGNATED MARKSMEN IN AFGHANISTAN

The concept of the Designated Marksmen or Squad Designated Marksman or Platoon Marksman was largely proven in Afghanistan. As we've previously noted, infantry fire teams were experiencing so-called weapon overmatch when confronting insurgent forces who were fielding 7.62x54mm PKM medium machine guns and SVD sniper rifles along with RPG rockets that could engage from 800 meters. Even the insurgents' 7.62x39mm AK47s outdistanced Coalition individual weapons, although they certainly suffered in any accuracy comparison.

U.S. Army Major Thomas P. Ehrhart convincingly argued the case in his 2009 monograph *Increasing Small Arms Lethality in Afghanistan: Taking Back the Infantry Half-Kilometer*.

> In general, the requirements for the infantry squad are that they have weapons capable of reliable incapacitation from close range to a distance of 500 meters. This capability does not exist in the current family of 5.56mm ammunition, either with military or with commercial off the-shelf ammunition, though efforts are underway to remedy the situation. Currently, the infantry squad does not have this capability unless its designated marksman is armed with a rifle of 7.62x51 caliber. Those armed with 5.56mm versions of the SDM-R are marginally effective and then dependent on shot placement in the small vital areas of the enemy for their effectiveness.[23]

The DM role was part of the solution to this issue, along with improvements in the design of the issue 5.56x45mm round, lighter-weight machine guns and longer-range grenade launchers. DMs also took up the slack in Afghanistan for overworked sniper teams, including conducting observation posts and providing some sniper overwatch tasks. The DM also provided his squad with the ability to PID targets at ranges exceeding the optics mounted on the issue assault rifles, an important consideration in a counterinsurgency war. By 2010, the U.S. Army was deploying two SDMs with each infantry squad in Afghanistan.

The British had been pressing their light support weapon, the L86A2, into service as a *de facto* DMR with some success in Iraq, but in Afghanistan, the weapon was still restricted by its 5.56x45mm caliber despite its longer barrel. By December 2009, however, the British had issued what is known as an urgent operational requirement for a dedicated 7.62x51mm DMR. In a testament to their resolve, the resulting weapon, the L129A1, was fielded within six months, in time for the 2010 fighting season. The British now have one marksman in every infantry section armed with the DMR.

The French Army in Afghanistan issued the 7.62x51mm HK417 as a Designated Marksman weapon. DMs were a platoon-level asset with two *tireur d'élite* (sharpshooters or marksmen) available under the platoon headquarters (initially there was only one sharpshooter authorized but by 2008 this had been increased to two purely based on Afghan experience). These marksmen were initially armed with the venerable but respected 7.62x51mm FR-F2, but HK417s were soon made available.

Indeed, French combat experience saw them adopt a number of non-standard platforms including the 5.56x45mm Heckler and Koch G36KV with underslung 40mm AG36 grenade launcher, as the issue FAMAS cannot easily accept an underslung grenade launcher. A typical French infantry section (*groupe de combat*) in Afghanistan, and now Mali, might see up to three HK417s, two Minimi light machine guns, a G36KV with AG36 and a number of standard-issue 5.56x45mm FAMAS.

The Danish Army did not have a Designated Marksman program, nor did they decide it was a capability they needed in Afghanistan. Instead their sniper teams were attached to infantry sections for specific tasks that in other armies would have been the provenance of the DM. The Canadians also failed to issue a DMR, again using their sniper teams attached down to platoon level. Australian forces began to field DMs as part of their infantry sections with firstly one man per section and later two, one in each fire team, equipped with a scoped 7.62x51mm SR-25 or HK417.

SOF SNIPERS IN AFGHANISTAN

During the early rotations in Afghanistan, the snipers from units such as the Rangers were heavily tasked with reconnaissance and surveillance missions. However, as the role of SOF became more and more dedicated to direct-action kill or capture missions, the snipers got back behind the trigger, supporting assault teams as they cleared target sites. Snipers would establish overwatch positions outside target compounds to guard against any unexpected surprises for the assault team.

Many SOF snipers operated far closer to the enemy than usual during these raiding operations. They would also commonly encounter large numbers of insurgents on or around an objective. The insurgents could also pop up from virtually anywhere. In one U.S. Green Beret operation in 2004, the Special Forces sniper element moved silently around a crop field, avoiding it to reduce noise and possible compromise before the assaulters were in position. Once in position directly outside the target compound, to provide supporting fire as the breach was conducted, a firefight broke out as a sentry spotted the team.

Moments later, the sniper team also began receiving fire from behind them. There had been several insurgents sleeping in the relative cool of the crop field who had now been awakened by the gunfire and were attempting to flank the sniper team. Several of the snipers turned to face the new threat while the others remained ready to engage targets in the compound. Technology and training won the fight with the insurgents in the crop field, as the snipers used their night vision goggles and the infrared lasers attached to their 5.56x45mm Mk12 rifles to identify and kill the attackers at ranges well under 50 meters.

Along with raiding, SOF snipers were also involved in a number of hostage rescues in Afghanistan. Many of these were committed by criminal gangs rather than insurgents, although the danger always lay in the risk of the kidnappers selling their hostage up the chain to the Taliban or al Qaeda. In September 2007, British SBS snipers in Lynx helicopters conducted an aerial vehicle interdiction (AVI) of an insurgent SUV carrying two Italian intelligence agents who were being held as hostages. The aerial snipers disabled the SUV with precision fire, allowing an SBS assault team to land nearby and take the insurgents under fire.

Indeed, this was an early example of the AVI technique being used in Afghanistan. It became an increasingly prominent tool as SOF understood its unique advantages from their war against al Qaeda in Iraq. Years later, during the controversial rescue attempt of British aid worker Linda Norgrove in 2010, SEAL Team 6 aerial snipers engaged and killed two insurgent sentries from the air with their suppressed 7.62x51mm HK417 rifles as the assault team landed on the objective.

The British SAS developed the art of aerial sniping literally to new heights during the Afghan war by deploying teams of snipers in Chinook twin rotor transport helicopters as aerial sniper teams. The snipers would punch out the Plexiglas windows and use tethered .50BMG Barretts and bolt-action .338 Lapua Timberwolfs to engage targets. Their precision fire was matched by the 7.62x51mm M134 miniguns manned by the helicopters' crew chiefs. These airborne sniper platforms were the next stage in evolution of tactics that began with the SAS and SBS using Lynx and Puma helicopters in Iraq for aerial sniping.

The American SOF also used similar techniques in Afghanistan. In fact, one Green Beret of the 3rd Special Forces Group, Sergeant First Class Stephan Johns, was awarded the Silver Star for his efforts as an aerial sniper. Covering a helicopter assault force breaching a targeted compound, Johns' own helicopter was forced down by enemy fire. The Green Beret defended the downed helicopter for 30 minutes by himself, dispatching nine Taliban who attempted to swarm the aircraft and its crew. Johns finally left the crash site after he ensured the downed crew were safe on the rescue helicopter that arrived to recover the aircrew and the Green Beret.

Later in the war, the U.S. Army Rangers began conducting overnight operations into the Taliban's backyard, infiltrating into their base areas and raiding compounds of intelligence interest. They would then withdraw to fortified compounds and await the inevitable insurgent onslaught. Snipers were very much an integral part of this effort. An example of the kinds of operations the Rangers were involved with during these Team Merrill missions included an operation into the Sangin Valley of Helmand. One platoon of Rangers, supported by a sniper team, a pair of reconnaissance teams and an anti-armor team equipped with the bunker-busting Carl Gustav recoilless rifle, was surrounded by some 300 insurgents.

Precision fire from the snipers along with close-air support helped keep the swarming insurgents at bay until an extraction could be planned. Coming in under fire, one of the MH-47 Chinook helicopters was shot down and forced to crash land. The sniper and reconnaissance teams established a protective cordon around the crash site, allowing the Rangers to extract on the remaining Chinooks. The snipers were some of the last to scramble on board under the covering fire of an orbiting AC-130 Spectre.

U.S. Marine special operations snipers also saw considerable service throughout the war in Afghanistan. Marine Corporal Franklin Simmons, a sniper and team leader with the Force Reconnaissance Platoon of 2nd Battalion, 7th Marines was awarded the Silver Star for his actions during a complex ambush in Farah Province in 2008. Simmons went to the aid of Marines pinned down behind an immobilized HMMWV (High Mobility Multipurpose Wheeled Vehicle or Humvee), maneuvering into a position on top of a berm where he effectively defiladed a number of insurgents in prepared firing positions, although exposing himself to enemy RPG and small-arms fire. Simmons singlehandedly dispatched 20 Taliban that day with his 7.62x51mm Mk11 Mod 0 during the eight-hour firefight.

SNIPING IN AFGHANISTAN

One of the most famous SOF operations of all time was conducted from Afghanistan, the killing of Usama bin Laden in the SEAL Team 6 mission known as Operation *Neptune's Spear*. It too involved at least one sniper. His primary role was to provide aerial overwatch as the assault teams fast roped onto the roof of the main building. After the crash of one of the Stealth Blackhawks, the sniper, armed with a suppressed 7.62x51mm HK417, landed and established an overwatch position to cover the assaulters as they breached into the building. He was also there to overwatch a two-man element that patrolled the perimeter with Cairo, their Malinois combat assault dog, to deter curious civilians or local police. It is not believed that the SEAL sniper fired any shots during the operation.

The Australian SASR heavily employed snipers during their capture or kill missions against insurgent leadership targets. Trooper, now Corporal, Mark Donaldson, an SASR operator who received the Victoria Cross for his rescue of a wounded interpreter amid a vicious Taliban ambush, was involved in numerous operations where SASR snipers played a central role. One mission was typical of many similar operations. A six-man SASR patrol, including two snipers, established a covert hide site to watch for two persons of interest, Taliban medium-value targets. After three days of watching, the pair emerged from a compound and the snipers confirmed that they were indeed the wanted individuals.

The patrol's JTAC vectored in an A-10A Warthog ground-attack aircraft, but incredibly both the bomb it dropped and a burst from its fearsome 30mm cannon missed the two Taliban, who ran back toward the cover of a building. One of the snipers narrowly missed and the pair retreated inside. The A-10A came around again and dropped a bomb on the house, flattening it and the insurgents. Another insurgent, apparently aware of the rules of engagement, managed to escape the aircraft and the SASR snipers by mingling with a group of women and children. With little choice, the patrol leader was forced to let the insurgent escape.

Donaldson's fellow SASR operator, Corporal Ben Roberts-Smith, was also awarded the Victoria Cross for his actions during a mission into a Taliban base area in the ShinazefValley in the Shah-Wali-Kot District of Kandahar Province. He was also an SOF sniper. The Australian Special Operations Task Group was conducting what was termed a "large scale disruption operation" to flush out Taliban leaders. A company of commandos was inserted into the area by helicopter to stir up the hornet's nest. Once signals intercepts (listening devices) located the Taliban leaders responding to the commandos' actions, the SAS would be sent in to capture or kill them.

Roberts-Smith described his responsibilities during the mission during an interview with the Australian War Memorial in 2012, and is worth quoting at length as he gives a unique insight into these kinds of operations. "My patrol provided aerial fire support from the fourth Blackhawk. We also constituted the reserve assault team and could help out on the ground if things got hairy."[24]

His Blackhawk was placed into a holding pattern at the end of the valley, although things "got hairy" there fast:

The peak was about 50 meters below us so when an insurgent with an RPG popped up from the rocks we saw him straight away. Our patrol commander, Sergeant P, screamed "RPG, right, three o'clock" but a sniper and I were already firing. We were sitting in the open door of the helicopter with our legs dangling over the side. The insurgent got the RPG away, and it passed just under my feet. At the same time, two more insurgents opened up with a PKM machine-gun and rounds began tearing through the cockpit and belly of the Blackhawk.

Roberts-Smith and his fellow sniper managed to engage and kill all three insurgents before landing to recover their weapons. With the firefight intensifying for the primary assault force, his reserve was committed, and after taking small-arms and PKM fire on approach, the Blackhawk managed to touch down amid a huge cloud of dust. His seven-man team split into two sub-elements and began to advance upon the insurgent positions. Ahead were at least three insurgent PKM machine gun teams.

After killing an insurgent who attempted to launch an RPG at him, Roberts-Smith and a colleague each threw a fragmentation grenade at the PKM positions to suppress them and he charged forward.

The nearest machine-gunner was 20 meters from me on the wall, with one AK47 guy on his right and, directly in front of me in a break in the wall, another on his left. Once I started moving, and this all seemed to happen very slowly, the two AK guys peeled back into the compound. I don't know whether it was poor training, panic or whether he found it too awkward to swing his PKM around, but the machine-gunner couldn't seem to get on to me until I reached the break in the wall. But I was already down on one knee and I shot him twice in the head.

He then managed to engage and kill another PKM gunner who was focused on firing off to his flank at the pinned down assault force. The operators cleared a building, killing another insurgent inside before the second team from the relief force was taken under fire by yet another PKM team up ahead of Roberts-Smith. A grenade sent the PKM team scurrying off into the tree line, but the SAS soldiers outflanked and killed them.

Then we came under spotter fire, that is, fire controlled by someone who had a good view of us. P asked the cut-off team to lay down fire just in front of us and we advanced behind it. Presently I could hear the spotter yelling instructions into his mobile, which led me to a bush-covered spider hole. I yanked the bushes off and killed him.

In all, the SAS unit of 25 operators and five attached Afghan SOF accounted for 70 insurgents during the operation. They later learned that ten of those were medium-value targets and the high-value target they had launched against had died of wounds received in the battle sometime later. Roberts-Smith used a 7.62x51mm Mk14 Mod 0 Enhanced Battle

Rifle borrowed from a Navy SEAL unit during his famous firefight. His Mk14 featured an EOTech holographic red dot sight with a swing-away magnifier that could be used for distance shots with a fixed 3-power magnification, apparently well suited to shooting from the air.

An Australian Commando sniper team from D Company, 2 Commando Regiment also currently holds the distinction of the world record for the longest distance sniper shot in the world. The operation involved two separate two-man sniper pairs in individual hide sites commanded by a section leader. The shot was taken in the Kajaki District of Helmand Province on April 2, 2012. At roughly 08:00 local time, the target they had been hunting presented himself. Once the target's identity was confirmed, the snipers and their spotters went to work developing a solution that would prove to be an extreme range shot, pushing the absolute boundaries of their .50BMG Barrett M82A1 antimateriel rifles and their Schmidt & Bender PM2 3-12 power optics.

The range was confirmed by laser at 2,815 meters (3,079 yards). The snipers were at an altitude of 1,160 meters or over 3,800 feet themselves. They adjusted their scopes and prepared to fire. The section leader used a command-initiated engagement, firing both sniper teams simultaneously. Several seconds later, the spotters watched through their Leupold 12-40 power spotting scopes as a .50BMG match bullet struck the target, killing him. His colleagues dragged the body swiftly away with no idea from where the round had originated.

The war in Afghanistan also produced its own version of the American Sniper Chris Kyle in the form of U.S. Ranger Sergeant Nicholas Irving. Irving became the most deadly Ranger sniper, with a total of 33 enemy kills after three and a half deployments to Afghanistan. His favored weapon was a 7.62x51mm SR-25 he nicknamed "Dirty Diana." Amusingly, he claimed his spotter's rifle, a .300 Winchester Magnum Mk13, "had no personality whatsoever."[25]

His longest-range kill was at some 883 yards (807 meters) with his SR-25, engaging an insurgent who was part of a Taliban ambush against a pinned-down Ranger platoon in Helmand. His first round struck the ground in front of the insurgent. Irving mentally made the adjustment to his hold-off and sent the second round into the man's chest just as he looked up. His closest range kill was less than 30 feet away (9 meters).

Irving was famously involved in a nine-hour ambush when he and his spotter were attached to a four-man Ranger Reconnaissance Company team. At times Irving and the Rangers were almost overrun by insurgents. He was also targeted by an insurgent sniper. According to Irving the enemy sniper was "really good," but a British SAS unit tracked down and killed the man several years later.

TRAINING THE AFGHANS

Snipers were also employed to build the capabilities of the Afghan National Army's fledgling sniper program. For much of the war, typical programs ran for a fortnight and aimed to impart the basics of marksmanship, range estimation and fieldcraft to Afghan students. The objective was to develop a cadre of Designated Marksmen who could further impart their newly developed skills to their infantry colleagues. It also allowed the trained individuals to provide some form of sniper overwatch during dismounted ANA patrols.

Afghan National Army snipers are today equipped with the 7.62x51mm M24 and the 7.62x54mm Dragunov SVD. They must complete a three-week training program instructed by NATO forces. Perhaps surprisingly, the ANA snipers are trained in some fairly advanced techniques including adjusting for windage and distance estimation.

CASUALTIES AND CONSEQUENCES

Along with the locale of the world's longest-range sniper kill, Kajaki in Helmand Province was also the site of the tragedy that befell members of the British Army's 3 Para Sniper Platoon in 2006, documented in the film

Kajaki (released in the U.S. as *Kilo Two Bravo*). On September 6, snipers from the platoon were moving into a location near the Kajaki Dam to provide sniper overwatch for a preplanned operation to hit a Taliban checkpoint. Graphically illustrating another of the deadly dangers that faced Coalition forces in Afghanistan, the young lance corporal in the lead stepped on a landmine that detonated, severing part of his leg.

The snipers were forced to attempt to clear a path through what they were increasingly sure was an old Soviet legacy minefield, and as they did so, another soldier stepped on a mine. The Soviets had been less than thorough with marking minefields and legacy mines were a constant threat. The MERT or Medical Emergency Response Team, an RAF Chinook equipped as something of a flying triage bay with a medical team on board, couldn't land near the casualty due to the mine danger. Nor was it equipped with a rescue winch like the American Pedro HH-60s that the patrol, and the Task Force commander, had requested.

As the Chinook attempted to land, it set off another mine with its down draft, wounding two more soldiers and one that would prove fatal to others. As a courageous medic attempted to reach the wounded, yet another man stepped on a mine. Eventually the Pedro helicopters arrived and the casualties were winched out, although one soldier, Corporal Mark Wright, sadly died on the aircraft. At the coronial inquest back in Britain, the coroner summed up by saying of the men: "You are courageous and utterly fearless. I have nothing but admiration for you and your fellow soldiers."

Other Coalition snipers also paid the ultimate price in Afghanistan. A Marine Scout Sniper Team Leader from 3rd Battalion, 5th Marines was posthumously awarded the Navy's highest award for valor, the Navy Cross, in 2011. Sergeant Matthew Abbate and his Scout Snipers were supporting an operation in Sangin's Green Zone when they were ambushed. Two Marines and their Navy Corpsman (the medic) were wounded by IEDs and he and his men were in fact mired in an IED belt.

The citation for his Navy Cross tells the story:

With total disregard for his own life, he sprinted forward through the minefield to draw enemy fire and rallied the dazed survivors. While fearlessly firing at the enemy from his exposed position, he directed fires of his Marines until they effectively suppressed the enemy, allowing life-saving aid to be rendered to the casualties.

Realizing that the casualties would die unless rapidly evacuated, Sergeant Abbate once again bravely exposed himself to enemy fire, rallied his Marines and led a counter attack that cleared the enemy from the landing zone, enabling the helicopters to evacuate the wounded.

After surviving the hellish ambush with barely a scratch, Abbate was killed in action just six weeks later.

The Afghan War was also not without its controversies, which occasionally involved snipers. An American sniper and his team leader, both U.S. Army Green Berets from the 3rd Special Forces Group, were charged with murder after they shot and killed a known and identified insurgent. The circumstances were that once the Green Berets had learned the insurgent's location in a nearby village, they established a support-by-fire position with a sniper while using their partnered Afghan National Police to call out the insurgent. When the man appeared, the sniper engaged and killed him.

While this may seem brutal, the individual had already been identified as a wanted enemy combatant in connection with both an IED and suicide bomber cell from Pakistan. The ROE in place at the time apparently required that the target be positively identified as an enemy combatant before they could be engaged. The American general responsible for the Afghan theater at the time considered the killing unjustified and attempted to have the men court martialed on murder charges. However, the charges were later dismissed. Such incidents illustrate that snipers must be very aware of their legal obligations with relation to the standing ROEs, lest they face criminal charges.

CHAPTER 4
SNIPING IN IRAQ

The city-based conflicts in Iraq, with their urban sprawl of mazelike streets and alleyways, offered a completely different experience from Afghanistan for the sniper. Operating in the cities required another skill set as snipers adapted to specific tactics for built-up areas, often concealed for days in covert observation posts. Iraq also saw different weapons being used and evolved as the engagement ranges reduced and the need for fast follow-on shots increased. The traditional bolt action was supplanted by a new generation of precision semiautomatic platforms.

The urban threat also evolved, with sniper pairs falling prey to swarm tactics by insurgents and necessitating the adoption of larger sniper security elements that often saw a sniper and his spotter accompanied by six to eight infantrymen to provide all-round defense. With a nod to Stalingrad, special operations snipers were also often deployed in countersniper operations, brought into an area to hunt a particular insurgent marksman. They also hunted the IED emplacers who planted roadside bombs in order to kill and maim civilian and soldier alike.

SNIPING DURING THE INVASION PHASE

Snipers were heavily committed in all conventional and SOF units during the initial invasion of Iraq. One of the first special operations of the war was largely successful thanks to the efforts of Ranger snipers. On April 1, 2003, the Rangers were assigned the seizure of Haditha Dam – known as Objective Lynx – to guard against Saddam Hussein blowing the dam wall and flooding the Euphrates Valley and advancing Coalition forces. The Ranger sniper sections proved their worth, engaging enemy spotters guiding in artillery and mortar fire. Two Ranger sniper pairs from the 3rd Battalion, 75th Ranger Regiment, equipped with the Mk11, M24 and M107, engaged targets out to 2,000 meters (2,187 yards).

A number of Iraqis were killed at ranges of up to 1,000 yards (914 meters) with the 7.62x51mm Mk11s (the Rangers' Company Sergeant Major, who had recently transferred from Delta Force, killed upwards of 20 Iraqis with his Mk11). The Ranger snipers also used their Forward Observer training to call in their attached 120mm mortars, and in one memorable engagement, more reminiscent of a video game, a round from one .50BMG M107 punched through an Iraqi soldier to ignite a propane tank behind him, killing two further enemy in the resulting explosion.

Special Forces ODAs assigned to Special Reconnaissance missions in the south, west and north of Iraq carried with them a range of sniper platforms, including the 7.62x51mm M24 and Mk11 and the .50BMG M107. These Green Berets, usually one or both of the weapons sergeants in an ODA, used their rifles to deadly effect. In one notable contact, a two-man reconnaissance team from ODA 525 was surrounded outside of the Iraqi town of Ar Rutba. As one operator called for emergency airstrikes against the encircling Fedayeen Saddam, the other Green Beret, a sniper school-trained weapons sergeant, engaged leadership and heavy weapons crews with his suppressed 5.56x45mm Mk12 Special Purpose Rifle, sowing enough confusion among the ranks of the militia to buy the airpower time to arrive overhead and begin dropping their ordnance.

Coalition SOF snipers were using their snipers to great effect in the western desert. Patrols from the Australian SASR struck on an unusual but highly effective method of convincing the Iraqi defenders to surrender first a factory complex and later a military airfield. Calling in limited airstrikes to destroy bunkers around the perimeter of the sites, they then requested low-level "show of force" flights over the Iraqi positions. Any Iraqis that weren't convinced to give themselves up were then subjected to precision warning shots from the Australians' 7.62x51mm SR-25s. These proved the final straw and both facilities were captured without further bloodshed.

On March 27, 2003, SEAL Team 6 and the Rangers launched an assault against a suspected chemical or biological weapons facility, the al Qadisiyah Research Center, otherwise known to the special operators as Objective Beaver. The Rangers established blocking positions around the target building to isolate its occupants from reinforcement, while the SEALs cleared through the building and conducted a search for evidence known as a Sensitive Site Exploitation or SSE.

Supporting the operations were a number of Little Bird helicopters of the 160th Special Operations Aviation Regiment, both the armed gunship variant known as the AH-6 and the unarmed troop carrier version, the MH-6. Perched on the external bench seats of two MH-6s flying low over the al Qadisiyah were a number of SEAL Team 6 snipers, one on either side of two of the tiny helicopters. Using suppressed 7.62x51mm SR-25s, the snipers engaged numerous enemy – both uniformed and black-clad Saddam Fedayeen militia – in and around the target site as they attempted to reach the Research Center.

The official history notes:

After the main force infiltrated, the MH-6 pilots were free to maneuver over the target and engage enemy combatants. (One of the pilots) spotted two Iraqi gunmen running from a driveway, each closely dragging a woman for protection. As (another pilot) positioned his sniper closer and lower, one gunman lost his grip on the woman, and the sniper immediately killed him. The second gunman backed into a concrete building and held the woman tightly. (The pilot) decided that it was more prudent to return to the target and reluctantly disengaged.

An enemy pick-up truck was spotted speeding toward one of the Ranger blocking positions, and machine gun fire from the Rangers caused it to stop. Two gunmen leapt from the vehicle: "The Rangers killed one instantly. The second gunman ran down an alley out of their sight, but well within range of a sniper's bullet from [another] aircraft; the man was shot dead within seconds." The effectiveness of the precision fire delivered from the MH-6s proved vital to allowing the SEALs to conduct their mission: "The Little Bird was like a fifty-foot mobile deer stand." We should remember that as a former JSOC (Joint Special Operations Command) officer with extensive combat experience in multiple theaters confirmed to the author, "shooting a moving target from a moving helicopter is one of the toughest shots there is. There's very few men who can do that." The snipers from SEAL Team 6 are certainly among those rare few.

British snipers were also busy in the south. In April 2003, the Irish Guards deployed their nine-man sniper section in and around Basra. Using their 7.62x51mm L96A1 rifles, the snipers were key to targeting Ba'ath Party officials and Fedayeen Saddam paramilitaries who were intent on keeping fighting, even as the conventional Iraqi military mostly faded away. The Irish Guards snipers' commanding officer noted at the time: "Our snipers are working in pairs, infiltrating the enemy's territory, to give us very good observation of what is going on inside Basra and to shoot the enemy as well when the opportunity arises." His men deployed around the outskirts of Basra in abandoned buildings where they constructed urban hides. "They don't kill huge numbers, but the psychological effect and the denial of freedom of movement of the enemy is vast. Our snipers have done really well."[1]

British mounted patrols in Warrior infantry fighting vehicles would drop the sniper teams off during routine fighting patrols into the city, a tactic that had been successful in Afghanistan and Northern Ireland before that. One sniper described the technique: "When a British raid goes into enemy territory, my sniper team goes along with the armored vehicles. The other soldiers fan out and sweep the area, then get back in their Warriors and leave. We stay put. The idea is that in all the confusion, the Iraqis wouldn't notice that not all the men who drove in drove back out again."[2]

One of the Irish Guards snipers described his experience of sniper operations in and around Basra:

Sometimes it's a bit hairy when we are getting to our position when there are rounds and mortars coming down around us. It's also a bit scary going into the buildings because they haven't been cleared and we don't know if they have left any booby traps for us. But once we are here they don't know where we are and it feels okay. We can report back on what's going on – to call in air strikes or direct artillery – and if they are within range of our rifles we will shoot them. I've killed two people for definite. When I got the first guy they brought up a second, and when I shot him they didn't send any more. I shot somebody else, but he went over a wall so I couldn't see what had happened to him.[3]

A Royal Marine sniper killed an Iraqi Army sniper at some 860 meters (940 yards) in high winds:

I knew I only had one shot and had to get the angle exactly right. It was hot and the wind was blowing strongly from left to right as we crept up to a vantage point ... I saw I had a clear shot at my man – he was in what he thought was a secure position but his head and chest were exposed. He was still wearing his green Iraqi uniform and was holding the rifle he'd been using to shoot at Marines.[4]

The sniper's spotter judged the sniper would have to hold off some 17 meters (18.5 yards) to the left and over 10 meters (11 yards) high due to the range, heat and strong crosswinds. Luckily the Iraqi sniper remained stationary as the sniper carefully lined up his shot. Squeezing the trigger of his 7.62x51mm L96A1, the round struck the man in the chest, killing him.

As the invasion wound down with the defeat of Saddam Hussein's army and paramilitary forces, many Coalition snipers and their units were sent home. Few had an inkling that a vicious insurgency that would last longer than World War II was brewing, an insurgency that would require the absolute precision of the sniper to defeat.

SNIPING IN THE IRAQI COUNTERINSURGENCY

"If a vehicle of armed men were to arrive, one shot from the Barrett .50 cal using armor-piercing rounds could stop it in its tracks, forcing the occupants out into the open without excessive collateral damage," explained a Spanish sniper.[5] Marine Battalion Commander Lieutenant Colonel Kenneth M. DeTreux argued that "Our Scout Sniper teams have a deterrent effect. That's not wishful thinking. The insurgents fear our snipers."[6]

As we've discussed in the previous chapter on Afghanistan, snipers make a very good tool in counterinsurgency. Both their kinetic and intelligence-gathering skills are extremely valuable in an environment teeming with civilian noncombatants, some of whom are insurgent sympathizers but most of whom are simply trying to live their lives with no great allegiance to any side in the conflict.

Officially the U.S. Army has the following to say about employing snipers in the urban environment:

Historically, snipers have had increased utility in urban areas. They can provide long- and short-range precision fires and can be used effectively to assist company- and platoon-level isolation efforts. Snipers also have provided precision fires during stability operations. Along with engaging assigned targets, snipers are a valuable asset to the commander for providing observation along movement routes and suppressive fires during an assault.

Sniping within what they term an "urban guerrilla environment" has its own set of unique complications.

1. "There is no forward edge of the battle area [FEBA] and therefore no 'no man's land' in which to operate. Snipers can therefore expect to operate in entirely hostile surroundings in most circumstances." This was particularly pronounced in Iraq, where the built-up nature of many of the operational areas and the questionable sympathies of the civilian population (and even members of the local security forces) meant that compromise was both highly probable and very dangerous to the operators involved.
2. "The enemy is covert, perfectly camouflaged among, and totally indistinguishable from, the everyday populace that surrounds him." While also a challenge in Afghanistan, and certainly in most counterinsurgency operations, this was also particularly pronounced in Iraq where gunmen could seemingly appear, and disappear, at will. It also made the work of the sniper that much more difficult as their targets would use the civilian population to mask their movements.
3. "In areas where confrontation between peacekeeping forces and the urban guerrillas takes place, the guerrilla dominates the ground entirely from the point of view of continued presence and observation. Every yard of ground is known to him; it is ground of his own choosing. Anything approximating a conventional stalk to and occupation of a hide is doomed to failure." In Iraq, likely hide sites were often known to the opposition, with the inherent risk that they may have been booby-trapped with an IED. The insurgent's knowledge of the lie of the land meant that snap ambushes or short-range sniper attacks were a constant threat to Coalition sniper teams.
4. "Although the sniper is not subject to the same difficult conditions as

he is in conventional war, he is subject to other pressures – these include not only legal and political restraints but also requirements to kill or wound without the motivational stimulus normally associated with the battlefield." Another important consideration in a counterinsurgency war is that snipers may be called upon to engage targets that are not directly at that moment threatening comrades or civilians, for instance where a high-value target or globally designated terrorist is eliminated by precision rifle fire or where a sniper is forced to engage civilians who are being used to transport ammunition, weapons or as spotters. In conventional war, the sniper is fighting against a known, uniformed enemy who fights, by and large, in a similar manner to himself. In counterinsurgency warfare, there are no such assurances.

5. "Normally in conventional war the sniper needs no clearance to fire his shot. In urban guerrilla warfare the sniper must make every effort possible to determine in each case the need to open fire, and that it constitutes reasonable/minimum force under the circumstances." This is a key factor and one that we will revisit at the end of the chapter when we look at how things can go wrong. Rules of engagement are often criticized, largely by the domestic civilian population, as being akin to fighting with one hand behind one's back. What is forgotten is that they also provide two important functions for the sniper.

Firstly they provide normally reasonably clear guidance on when they are allowed to shoot. This may seem obvious, but in a counterinsurgency environment it is often far from clear what constitutes a hostile act. ROEs spell this out for the soldier. Secondly, they provide guidance on, perhaps more importantly, when not to shoot. When conducting a "war amongst the people," as aptly termed by General Rupert Smith, engaging the wrong person, or more problematically the right person at the wrong time, leads only to further support for the insurgency.

Indiscriminate killing does nothing to win the proverbial hearts and minds of a suspicious, if not downright hostile, civilian populace.

Snipers allow for measured and precise application of force against identified enemy personnel in the counterinsurgency environment. U.S. Army sniper Sergeant Randall Davis explained in a 2003 *Army Times* interview: "I just thought it was a very smart way to fight a war – very lethal, very precise. This way I know I'm not shooting civilians. Every shot you take, you know exactly where the bullet is going."[7]

In the predominantly urban environment of Iraq, Coalition snipers began to modify their shooting techniques to match that environment and what were known as "limited exposure targets." In an urban setting, many targets will be fleeting and if a shot is missed, the target will likely melt away down alleys and rat-runs. To counter this, many sniper teams began to employ multiple rifles to engage a single target.

Also related to the constantly moving nature of the urban insurgent, snipers began to aim for center of body mass rather than going for the more difficult headshot, as this increased the chance of a successful hit. If the insurgent was still showing hostile intent after the center of body mass hit, the sniper could reengage. Headshots, being a smaller target, also naturally take longer for a sniper to line up. Maximizing engagement speed is important when the sniper is confronted with multiple insurgent targets.

Semiautomatic sniper rifles also came into their own in Iraq. As we have seen, there is traditionally some resistance to the employment of semiautomatic platforms, as many snipers still perceive them to be less accurate and less reliable than the traditional bolt action. Modern semiautomatic designs like the 7.62x51mm SR-25, HK417 or Mk11 Mod 0 are very accurate platforms.

It is only really at extended ranges or in a bench rest (a device that locks a rifle in a vice that is used to test inherent accuracy) where the increased accuracy of the bolt-action design becomes at all evident. In Iraq, most of the sniper shots were at urban combat ranges of two to three hundred meters, well within the range to make the most of the capabilities of the semiautomatic. Having said that, there was also scope for the occasional extended distance shot, such as when Sergeant Brian

Kremer of 2nd Battalion, 75th Ranger Regiment made a record 2,515-yard (2,230-meter) shot in March 2004 using the .50BMG M107.

Marine sniper Sergeant Tim La Sage of 2nd Battalion, 5th Marine Regiment interviewed in 2015 explained his preferences in Iraq. "Man, the M40A1 was used forever. Bolt action is the best, but in a city it can be complicated. In Ramadi, I thought the semi-auto SR25 was great; it allows you to rapidly fire from a 20-round magazine. That's hard to beat in a city."[8] U.S. Army sniper Sergeant Randall Davis used only semiautomatic weapons, including an optically equipped 5.56x45mm M4 and a 7.62x51mm M14 along with the .50BMG M107 SASS to make his shots in Iraq.

Staff Sergeant Dillard Johnson served multiple tours in Iraq, eventually chalking up 121 insurgent kills. Rather than the more traditional M24, he used the 7.62x51mm Mk14 Mod 0 Enhanced Battle Rifle, the modernized version of the M14 mounting a Bushnell Elite mildot optic. "I could shoot two or three at a time when they were trying to plant an IED, if I was fast. One time we set up in a building and left the lights on. I shot five with the lights on and six after we turned them off,"[9] he explained.

The primary advantage of the semiautomatic in Iraq was obviously the capability to deliver a fast follow-up shot or to engage multiple targets at a speed that was simply unavailable to even the most experienced shooter using a bolt action. Snipers in Iraq would often confront multiple insurgents either preparing an ambush or emplacing an IED, mortar or unguided Chinese 107mm rocket. The semiautomatic platform allowed these snipers to quickly engage all of the identified insurgents before they could escape. Snipers are drilled to engage "near to far" in terms of target priority, taking out the closest targets first and working their way back. The semiautomatic gives them the capability to quickly do so.

Another advantage of the semiautomatic was that it could be pressed into service to clear rooms or conduct more regular infantry street fighting, negating the need for the sniper to carry a second weapon like an M4. Instead, the semiautomatic could function as an infantry rifle if the situation dictated. This is portrayed in the film *American Sniper*, where Chris Kyle assists a squad of Marines clearing houses in Fallujah with his SEAL-issued semiautomatic Mk11 Mod 0.

Silver Star recipient Staff Sergeant Dillard Johnson made a telling comment on the nature of the sniper war in his excellent account of his multiple tours in Iraq, *Carnivore*. "In Afghanistan they were doing traditional sniping, sneaking and peeking and hitting skittish Taliban from way off, but our environment was a lot closer to what police snipers have to deal with." Even though the ranges were typically reduced in comparison to Afghanistan, snipers still found themselves engaging at respectable distances. The "Widow Makers," a sniper team from the 5th Squadron, 73rd Cavalry Regiment, clocked up their first kill on an insurgent RPG team at a lased distance of 625 meters using the .50BMG M107 SASS with a Leupold Mk4 scope.

The Iraqi insurgents began to develop and use self-taught countersniper drills. Many Coalition snipers who had reported high kill tallies in earlier tours were seeing only a dozen or fewer by a whole sniper platoon by 2005 and 2006, the height of the insurgency. With conventional forces rotating through as ground-holding units typically in the same AO (Area of Operation), and based from the same PBs and FOBs, the insurgents got to understand the best locations for sniper hides and observation posts. As most of the insurgents were locals, they also knew the best covered avenues of approach that would shield them from a sniper's sights and used these to ambush and emplace IEDs.

The insurgents in Iraq also learned to take advantage of the ROEs that Coalition forces were required to follow. During much of the Iraq war, an insurgent needed to be visibly armed to be engaged. The insurgents soon understood this requirement and rarely moved within sight of Coalition snipers carrying any sort of weapon. Instead, weapons were either cached or carried concealed by civilians or in vehicles. In Fallujah, insurgents even resorted to using Red Crescent ambulances to move weapons and ammunition, and insurgents, around the battlefield.

The father of modern sniping, Major John L. Plaster, explained in a 2010 interview the predicament facing Coalition forces in Iraq:

> The terrorist snipers' tactics proved as extreme as their philosophy. Unencumbered by international law, insurgent snipers hid themselves in civilian clothes, took cover behind human shields, fired from mosques and escaped in ambulances. With a seemingly endless supply of sniper rifles and thousands of American targets, these hit-and-run terrorists could strike anywhere. American quick-reaction forces rushed to such scenes, but most often the enemy snipers simply dumped their rifles and blended into the neighborhood. They seemed nearly impossible to kill or capture.[10]

Somewhat surprisingly, the annals of Coalition sniper tales from Iraq also include a large number of mentions of Iraqi civilians actively assisting the Western sniper teams. Sometimes this included feeding the snipers (for which the civilians were recompensed from unit funds) or hiring out their roof or front room as a hide site (again for which they were generously paid). They also provided tip-offs about the location of IEDs or the whereabouts of insurgents. Particularly as al Qaeda in Iraq ramped up their murderous campaign to incite a civil war, many Iraqis would choose to support the Coalition rather than the foreign jihadists.

SNIPER OVERWATCH MISSIONS

Like in Afghanistan, Coalition snipers would often provide sniper overwatch, offering a largely invisible ring of protection for their colleagues. Both mounted and dismounted infantry patrols would receive sniper overwatch wherever practicable. These patrols could be conducting a range of missions: presence patrols that aimed to reassure the civilian population while keeping the insurgents off their feet; area-denial patrolling to lock down a specific neighborhood to insurgents; vehicle check points; raids targeting particular persons of interest; or cordon and search operations looking for weapons or IED caches.

The overwatch role also meant that snipers would often be able to identify suspicious items in the path of a Coalition patrol. The insurgents would hide their IEDs in literally anything, including the corpses of dead animals. Snipers also watched for the telltale blast of the RPG. The RPGs favored by the Iraqi insurgents had one fatal flaw: their back blast would lead directly back to their point of origin. Snipers would watch for this and engage once the RPG operator was identified.

Snipers also provided protection for the Vehicle Check Points or VCPs that were a fact of life for most Coalition soldiers deployed outside the wire. Their aim was to restrict the freedom of movement of insurgents and terrorists by deploying temporary or snap VCPs (or flash traffic control points, TCPs, as they were known in the U.S. military) that would stop cars and search any suspected of carrying insurgents.

Although the work was viewed as dull and monotonous by most soldiers, it was essential and had a significant effect on insurgent activity. A program of snap VCPs in a particular area could actually shape insurgent actions. It could also often turn up concealed weapons or explosives as unwary insurgents were caught in the net. Despite the monotony, VCPs/TCPs could also be deadly, as the soldiers would never know whether the approaching car was a suicide VBIED (Vehicle-Borne IED) or simply an erratic Iraqi driver. Marine Scout Sniper Sergeant Matthew Mardan said in a 2007 interview of his role as a sniper: "People think that's dangerous, and it is, but I would do that any day of the week rather than be a Marine sitting on a fucking checkpoint looking at cars."[11]

These VCPs/TCPs were often overwatched by a sniper team. The snipers had a number of responsibilities. Primarily they were watching out for insurgents trying to ambush or snipe at the VCP/TCP. They were also looking for car bombs. Distinctive signs included an obviously overloaded vehicle sitting well down on its axles, a nervous or agitated driver, a driver who appeared to be chanting, or a vehicle that ignored all directions and warnings.

If a sniper spotted a suspect vehicle, the unit responsible for the VCP/TCP would be immediately notified and they would react according to what were known as Escalation of Force (EOF) procedures. These EOF levels included the use of signal flares, pen flares, visible-spectrum targeting lasers, spotlights and even flashbang grenades, along with audio warnings. If the suspect vehicle continued heading toward the VCP/TCP, weapons would be pointed directly at the vehicle followed by aimed warning shots in front of the vehicle.

If the suspect vehicle still continued to approach, shots would be fired into the engine block and finally escalated to directly engaging the driver. Snipers were often an integral part of the EOF process as their superior optics could discern hostile intent at far greater ranges than the combat optics mounted on most Coalition Forces' individual weapons (many of these, like the common Aimpoint, provided no magnification). Dependent upon the rules of engagement in place at the time, the sniper might even engage the vehicle or the driver if hostile intent could be confirmed.

Person-Borne IEDs or PBIEDs, more commonly referred to as suicide bombers, were also another constant threat to VCPs/TCPs. PBIEDs held an added danger in that even if the suicide bomber was successfully identified and engaged before he or she could detonate their bomb vest, the device could very well have been equipped with a "fail-safe" initiator, allowing a third party to detonate the bomb from a distance.

Snipers would watch for lone individuals approaching the VCP/TCP who were, like suicide VBIED drivers, sweating profusely, were chanting or praying and who might have been wearing an unseasonal or uncharacteristic amount of clothing that could conceal a device. Again, their optics aided them in this task. Should an individual be confirmed as a suicide bomber and fail to respond to verbal challenges or warning shots, the sniper was ideally placed to end the confrontation with a precision shot to the fatal T, ensuring the round did not inadvertently strike a detonator or explosive charge.

Snipers would also watch for IED emplacers attempting to deploy concealed IEDs to catch the VCP unit upon their withdrawal from the site. Once the snipers spotted a suspect device or suspect individuals, they would radio the patrol, who would immediately establish a security cordon around the suspected device while scanning for threat indicators. These might include fighting-age males milling about or arriving in civilian cars, a video cameraman (used to record the blast for propaganda purposes – also a common occurrence with insurgent sniper attacks as we shall later see) or suspicious individuals with cell (mobile) phones or radios who may have been acting as the triggerman for a remote-controlled IED or RCIED. Once the patrol had the situation locked down, EOD would be requested to investigate and if necessary make safe the device.

Even after the arrival of the EOD operators, the snipers had to maintain their watch. On a number of occasions, EOD teams were lured to fake devices so that they could then be engaged by insurgent marksmen. EOD were an increasingly priority target for the insurgents in Iraq. Snipers would also have to remain vigilant against insurgents attempting to trigger a secondary device to catch the EOD team or anyone with a video camera who might be filming the EOD units' tactics and techniques for distribution to the insurgency.

U.S. Army Sergeant Randall Davis from B Company, 5th Battalion, 20th Infantry Regiment, was interviewed soon after his 2003 tour. Davis deployed at the time with only his spotter in a two-man sniper element without additional security. As the insurgency developed its tactics, small teams like this became a significant risk, and after the killing of several Marine sniper teams, snipers were no longer leaving the wire in such small elements. In 2003 the enemy had also yet to adopt the IED as their primary weapon in the asymmetrical conflict and would still conduct traditional ambushes, often using bait to draw a Coalition unit into the kill zone. The British Army referred to such tactics as the "come-on ambush."

Davis' unit was ambushed in exactly this way by a numerically superior force that managed to channel the Americans into a dense built-up area. Davis provided overwatch with his M4, killing seven of the 11 insurgents killed in the contact. His use of the comparatively short-ranged M4 belies the fact that the majority of his targets were less than 300 meters (328 yards), although he later made at least one kill at 750 meters (820 yards) with the .50BMG M107.

Snipers were also used in defensive overwatch, deployed at PBs and FOBs to guard against insurgent attack. Like overwatch of a VCP, they needed to be on the lookout for all manner of assailants. When first deploying on such an operation, the snipers would develop range cards based on likely targets and landmarks, and begin to understand the local dynamics of the area so that they would be able to discern when something out of the ordinary was occurring.

British sniper Sergeant Dan Mills' superb account of his time as a sniper defending CIMIC (Civil-Military Cooperation) House in al Amarah, southern Iraq, *Sniper One*, has justifiably become a classic account of the modern sniper at work. The book describes Mills' experiences as the sniper platoon leader with Y (Support) Company, 1st Battalion, The Princess of Wales' Royal Regiment, during the siege of CIMIC House in 2004. His opponents were the tough and determined Shia militias loyal to the cleric Muqtada al Sadr, covertly supported by the Iranians.

His men were employed to protect the CIMIC building, a visible sign to the militias of the new Iraqi government. They were attacked daily by RPGs, mortars, snipers and small-arms fire. Patrols into the surrounding suburbs always required an armored QRF on standby. The British were initially constrained by ROEs that forbade them to engage insurgents until they were in the actual commission of a hostile act. Mills recounts spotting a truck carrying an 82mm mortar in its bed soon after CIMIC had received mortar fire. As the men were not actively engaged in hostilities, the snipers could not fire.

When they could, Mills' snipers would establish what were known as reactive OPs, ambushes by any other name, triggered from covert hide sites. As the months wore on, the attacks against the base became more brazen and concerted. Resupplies were brought in on Warrior Infantry Fighting Vehicles. Finally, a number of full-scale assaults were launched by hundreds of insurgents, which the defenders of the isolated outpost likened to something of a modern-day Rourke's Drift.

The sniper platoon racked up over 200 confirmed kills, and according to Mills, double that as unconfirmed kills where the bodies were dragged away or badly wounded were evacuated in civilian cars. By the time his unit rotated out from CIMIC, they had received "a total of 595 incoming mortar rounds … 57 separate RPG attacks and 5 barrages from 107mm rockets. We'd fought 25 different firefights out in the city itself, and repelled 86 enemy ground attacks on CIMIC itself."[12] Upwards of 33,000 rounds were fired back at the insurgents, a solid percentage from the sniper platoon.

The situation in southern Iraq remained dangerous. In 2007 in Basra, British snipers were supporting the Staffordshire Regiment providing sniper overwatch. The sniper pair were under fire themselves when the lead sniper was struck in the chest by a round that was luckily saved by the plate in his body armor. Moments later, an insurgent RPG team that had been firing at British forces for several hours broke cover. The sniper, although wounded, exposed himself to further enemy fire to engage the RPG team. With his L96A1 rifle, he shot and killed all three in rapid succession.

The same sniper experienced one of the saddest elements of the vicious insurgency in southern Iraq later that day: the use of children as spotters. "The two of us stared across at each other," the sniper said in a 2011 interview. "He was scared, shaking. Every time he wanted to go back in through the curtain, they pushed him out. I turned the weapon on him, then to the side and fired two meters away. He just chucked the phone and ran. I have never regretted it. I have two little cousins the same age."[13]

By 2008, the British were experimenting with aerial overwatch in southern Iraq with a unit from the RAF Regiment. The snipers undertook a nine-week training course to enable the snipers to fire effectively from the air mounted in Lynx and Merlin helicopters. As we have previously noted, such shooting is one of the most difficult sniper disciplines, but gives tremendous advantages to the sniper. "Operating from the helicopters offers us great observation and a different perspective to that on the ground. We are able to assist in securing areas rapidly and are able to engage targets at greater ranges," explained one RAF Regiment sniper.[14]

"We act as top cover to provide protection during many tasks. The tasks since deploying to Op Telic have included covering U.S. Blackhawks as they evacuate casualties from Basra Palace, covering UK Merlin helicopters inserting troops. We have also been employed in covering urban areas likely to be used to launch rockets against the (combat outpost)." The aerial snipers used the .338 Lapua L115A3, giving them increased range over the 7.62x51mm L96A1. Interestingly, their SOF brethren routinely use semiautomatic sniper rifles when conducting aerial shooting although this option was not officially available to conventional UK forces until the 2009 adoption of the L129A1.

COUNTER-IED MISSIONS

As the use of IEDs in Iraq increased, so too did the use of snipers and sniper rifles to counter the threat, be it counter-IED ambushes against the insurgents emplacing the devices or EOD deploying .50-caliber antimateriel rifles to destroy suspected devices at extended range. IEDs and VBIEDs, both static and mobile, were often used to initiate so-called complex ambushes of small arms and RPG fire. The IED or car bomb would be triggered first to stun and try to cause a casualty before the ambush itself was sprung.

Snipers were regularly deployed to overwatch known and popular IED laying points. Perhaps not surprisingly, the insurgents were creatures of habit and would reuse locations that had worked successfully for them in the past. A key role for an infantry battalion's intelligence shop was to ensure IED locations were tracked, and this information was passed on to the unit that would eventually relieve them. In this manner, a database of IED emplacement points could be developed and used to task intelligence-driven sniper missions. A British sniper explained one counter-IED mission: "The three Iraqis I killed were putting out antitank mines. I was taking a life to save lives. Each man would have caused casualties to my comrades if I hadn't have fired."[15]

A key counter-IED technique is pattern-of-life surveillance conducted by sniper teams in covert hides or observation posts. These operations would often see the snipers deployed for up to three days watching a specific area. Passively observing an area of interest allows the snipers to begin to understand the local dynamics of an area: who is local, who is not; what time certain events occur; what the civilian reaction is to the arrival of insurgents or a Coalition patrol. It also allows them to develop ground knowledge of likely IED emplacement sites like roadside culverts. Snipers experienced in surveilling an area of interest will be able to pick up on minute changes to behavior that may indicate an insurgent or IED presence.

Combat troops in both Afghanistan and Iraq referred to such insights as "atmospherics" or "combat indicators." The wrong sort of atmospherics, like individuals with video cameras or Motorola walkie-talkies or an area devoid of civilian life, tends to indicate an imminent IED strike or ambush. In Iraq, the locals inevitably knew who the foreign fighters were and who was a local member of one of the insurgent factions.

In Iraq, the huge number of armed militias and paramilitary organizations made target identification sometimes difficult, particularly after the advent of the Sons of Iraq program and the Anbar Awakening, where former insurgents pledged to fight against al Qaeda and foreign terrorist elements. A fighting-age male carrying an AK47 was sometimes not enough to definitively identify the target as hostile. A key element of counterinsurgency strategy is to attempt to convince local civilians to tell Coalition forces about individuals involved in the insurgency or the location of an IED hidden in their neighborhood, so shooting a member of the security force for the local imam was counterproductive. Thankfully, many insurgents, somewhat foolishly, began wearing balaclava masks that did tend to tip off Coalition snipers to their hostile intent.

Patient sniper surveillance was proven to be one of the key measures against the insurgent IED emplacers. U.S. Marine snipers from 2nd Battalion, 2nd Marines were employing such tactics as early as 2004, with two sniper teams deploying in concealed hides with overwatch of each other. In August of that year, one of the Marine sniper elements witnessed a highly suspicious scene. "We saw a tractor driving down the road with its headlights off ... and then a van that flashed its lights at the tractor – that alerted us and when the van pulled over next to the vehicle parked on the side of the road we knew something was up," explained a Navy Corpsman attached to the team. "When I saw two Iraqis get out of the van and begin to feed a spool of wire into an abandoned van, I thought 'this is too good to be true.'"[16]

Not all sniper shoots require the long gun. Due to the close range of the encounter, some 75 yards (68 meters), and the likelihood of fast follow-up shots being needed, the Marines instead employed their 5.56x45mm M16A4 rifles equipped with the 4 power ACOG optic and their under-barrel-mounted 40mm M203 grenade launchers. The contact was initiated by the snipers with a number of 40mm grenades before the snipers left their hides and advanced on the insurgents. They detained one insurgent, although his compatriots had already dragged away their dead and wounded.

One of the most famous U.S. sniper teams in Iraq conducted many counter-IED missions. It was in Ramadi in 2005 where Coalition forces were experiencing up to 100 IED incidents per week. First Sergeant

James Gilliland led the legendary "Shadow Team," a ten-man U.S. Army sniper element who managed some 276 kills in six months, with three to four insurgents being killed on average every single day. Gilliland's sniper section established a hide site on the roof of the tallest building in the area, a former hotel overlooking Route Michigan, the main highway through Ramadi, and went to work on the bombers.

Gilliland's team's location became so well known among the insurgents that they eventually would not journey any closer than 800 meters in line of sight of the hide location, as they feared being shot. This achieved one of the snipers' primary objectives – to reduce insurgent freedom of movement in the city. It also kept the vital Route Michigan clear of IEDs and severely depleted the ranks of the local insurgents. With the risk of sudden death at the hands of an unseen sniper, volunteers willing to lay IEDs became much scarcer.

One of the longest-range shots taken in Iraq also occurred on a counter-IED overwatch mission. In the town of Lutayfiyah in 2004, a U.S. Marine Scout Sniper pair was deployed on the top of an industrial oil storage tank providing overwatch for a Marine dismounted patrol some 914 meters (1,000 yards) away. Marine Staff Sergeant Steve Reichert had dragged the heavy .50BMG Barrett M107 SASR up onto the top of the tank and was peering through the scope, monitoring the patrol's progress.

Reichert spotted a suspected IED concealed in the body of a dead animal by the side of the road and advised the patrol commander, who immediately halted his patrol and established a security cordon around the device. Moments later, the first RPG streaked through the air and the Marines were engaged by a large force of insurgents. It became apparent the IED was likely a command-wire device and had been meant as the initiator of the insurgent ambush. Reichert had spoilt that plan.

The Marine sniper identified an insurgent machine gun team at 1,600 meters (1,749 yards) and carefully set up the shot. The sniper fired his Barrett through the brick wall the insurgents were sheltering behind for cover. The Raufoss Mk211 rounds with their tungsten steel penetrators made short work of the bricks, coating the wall behind the insurgent position in blood. He continued firing until all three were dead. No insurgents ran to pick up the PKM after that.

Sergeant Tim La Sage, another Marine Scout-Sniper deployed to Ramadi, described the war against the IED emplacers.

We were watching a main avenue of approach, observing a roundabout in a crowded marketplace. I was eight stories up, about 670 yards from a spot where the enemy was laying IEDs. There was a guy on a moped who kept dropping someone off to lay them. Every time we'd take out the guy on the ground, the guy on the moped would drive back with another one. We were letting the guy on the moped live because he was giving us people [to target]."[17]

La Sage also engaged an entirely different form of IED later in his tour, when he engaged an apparent suicide VBIED or car bomb. "I'm about 700 yards away. I am shooting from a tripod, standing. I see the vehicle is heading for the convoy. It's closing fast. I had to make a decision. So I take the shot. I can see red through the windshield, the vehicle straightens and hits the curb. It was a very lucky shot. When you shoot through glass you never know which way the bullet will go." This shoot was uncannily similar to that portrayed in the film *American Sniper*. A HMMWV-mounted patrol dispatched to clear the suspect vehicle found it was packed with explosives, and EOD was called to render the car bomb safe.

As the war progressed, so did the skills and inventiveness of the bombers. IEDs were soon placed against roadside guardrails, in storm water culverts under the road, even tied to trees in the median strip. In the end, an IED ended La Sage's tour as he and his sniper team approached a building to set up sniper overwatch for another unit raiding a terrorist safe house. As they crossed a street, the command-detonated IED was triggered, killing two of La Sage's comrades and grievously wounding the young Marine.

The 'top gun' of the U.S. Army in Iraq, a sniper team leader with the 4th Infantry Division, also regularly intercepted insurgent IED teams. In fact many of U.S. Army Staff Sergeant Timothy Kellner's 139 confirmed kills were against insurgents caught in the act of deploying IEDs. In fact, Kellner shot many more than the 139 he is credited with, but the bodies of the insurgents were dragged away before the kills could be confirmed. Some believe he may have increased his tally to over 300. With the variable ranges encountered in Iraq, Kellner also used both the M24 sniper rifle and the Mk14 Enhanced Battle Rifle (EBR). The EBR allowed Kellner fast follow-up shots as the IED emplacers would typically move quickly away from the site of the device and attempt to blend in with the civilian populace.

COUNTERSNIPER MISSIONS

The U.S. Army provides the following guidance for its soldiers who encounter an irregular sniper in an urban environment like Iraq:

In an urban area, the direction of enemy fire, especially from a single rifle shot, is often difficult to determine. If a unit can determine the general location of a sniper, it should return suppressive fire while maneuvering to engage the sniper from close range. This is not always successful because a well-trained sniper often has a route of withdrawal already chosen. Massive return of fire and immediate maneuver can be effective against short-range sniper fires, if the ROE permit this response. In high-intensity urban combat, they are often the best immediate responses. Exploding fragmentation rounds, such as 40-mm grenades from the M203 grenade launcher, are the most effective suppressors.

Soldiers and Marines deployed to Iraq during Operation *Iraqi Freedom* would use suppressive fire wherever possible, both to dissuade an insurgent sniper from continuing to fire and to fix him in place to allow his position to be flanked. As U.S. Army doctrine openly states, however, locating the source of the enemy fire, and thus the sniper, can be a very difficult task indeed. As we will soon cover, a number of technological means were developed to assist; however, the old maxim remains true that the best counter to a sniper is another sniper.

Army Sergeant Dillard Johnson was himself shot by an enemy sniper in Iraq, though thankfully it struck his Interceptor body armor. He was also involved in that rarest of confrontations, a sniper duel. In Salman Pak in December 2005, an insurgent sniper using a 7.62x54mm Romanian PSL narrowly missed Johnson and a fellow sniper. Johnson killed the insurgent sniper with the second round from his Mk14 EBR, his longest-range kill at over 850 meters (929 yards).

His partner, Staff Sergeant Jared Kennedy, behind the big .50BMG M107 Barrett, engaged and killed the insurgent sniper's spotter with a headshot. Incredibly, the insurgent spotter, also armed with a Romanian PSL, had fired at exactly the same moment as Johnson's partner fired the Barrett. The 7.62x54mm round had grazed Kennedy's arm, causing a minor but painful wound. U.S. Army sniper Sergeant Randall Davis was also involved in a sniper duel in Iraq, commenting after he killed the sniper (the eighth kill of his deployment): "We had been engaged by snipers here before, so I was hoping it was the same guy – it's kind of a professional insult to get shot at by another sniper."[18]

In another well-known incident, a U.S. Army infantry squad had taken up positions on the roof of a building to provide security during an operation. Army sniper James Gilliland's *Shadow Team* snipers could see the infantrymen profiling themselves against the sky in a practice known as skylining. After warning them repeatedly over the radio, the inevitable happened and an Iraqi insurgent sniper positioned in a nearby hospital managed to shoot one of the soldiers. The snipers of *Shadow Team* immediately began searching through their optics for the gunman.

Gilliland himself identified the insurgent sniper at the hospital, hiding in a fourth-story bay window and armed with an SVD-type rifle.

Gilliland's laser rangefinder had earlier reported the hospital as some 1,250 meters (1,367 yards) away. Gilliland, armed with the issue bolt–action 7.62x51mm M24 with Leupold Ultra M3A 10 power optic, knew that the insurgent was beyond the 800-meter (875-yard) effective range of his rifle.

He dialed the Leupold out to 1,000 yards (914 meters) and to compensate for bullet drop, he held off three mils above the target (the equivalent of 12 feet or 3.6 meters) and 8 feet (2.4 meters) left of the target to account for the effects of the wind. At this range Gilliland was only hoping to suppress the enemy sniper to stop him from further engaging the infantry squad. Incredibly, his shot struck the insurgent dead in the chest, rendering him immediately *hors de combat*. At the time Gilliland's shot was the longest 7.62x51mm sniper shot in history and is still the longest known 7.62x51mm shot in Iraq.

The Coalition responded to insurgent sniping in a number of ways. Ballistic helmets were improved and body armor coverage increased, including the introduction of additional trauma plates covering the side of the body. Gunners in MRAPs and HMMWVs were provided with additional body armor components covering their neck and upper arms, while the British introduced the Kestrel, a set of heavy armor designed to protect vehicle gunners from sniper fire and the blast and shrapnel from IEDs.

The design of vehicles also changed, with protection for the gunner and crew finally becoming a priority. Ballistic glass was fitted and a successive range of improvements were made to gunner's turrets, including the Gunner Protection Kit (little more than a gunshield) and the later Objective Gunner Protection Kit that saw a full turret with ballistic glass fitted. Eventually turret-mounted gunners were replaced with Remote Weapon System mounts that allowed the heavy weapon to be fired from within the comparative safety of the vehicle.

An acoustic sniper detection system was also fielded that operated on a similar principle to that of countermortar radars. The system was named Boomerang and gave an audible warning and target direction when a sniper fired within range of the system. Boomerangs were deployed both on vehicle mounts and installed at patrol bases and forward operating bases. In areas that were particularly plagued by insurgent snipers, the Boomerang would be complemented by a Coalition sniper team that would use the data provided to target and eliminate the enemy snipers.

Former British Army Sniper Master, Mark Spicer, argues that "the true counter sniper option is to have used tactics and drills that severely restrict if not negate the ability of your enemy to deploy snipers against you in the first place."[19] When out on dismounted patrols, Coalition soldiers were trained to continually move to avoid being engaged by an insurgent sniper, as their skills rarely stretched to hitting a moving target. They also practiced a technique known as parallel patrolling that saw squads patrolling at different speeds on adjacent roads: this may not stop the sniper from firing, but it significantly lowered the chances of him escaping as he could not be sure exactly where the Coalition troops were located. This tactic would have inevitably led to some insurgents deciding that discretion was indeed the better part of valor.

The Marines experienced a large concentration of insurgent snipers during the August 2004 Battle of Najaf. During one 24-hour period, the Marines lost two men to sniper fire. They responded with their own Scout Snipers and attached Coalition SOF, killing 60 insurgents. The official history details another countersniper technique:

In one unconventional antisniper tactic, Captain Sotire requested tank support. The militia snipers, firing from positions on an extended overhang attached to building 61, Company A's base, continued to give the unit's own attached snipers a difficult time. Upon request, one of Lieutenant Thomas' tanks sent a 120mm round into the enemy building and brought down the entire structure, killing all the militia snipers who had nested there. Alpha's own snipers did not sustain any injuries.

COUNTER-IDF MISSIONS

Another mission sniper teams would be tasked with was the counter-IDF or Indirect Fire mission. In Iraq this was primarily against enemy 82mm mortar positions, although the insurgents also fired unguided Chinese 107mm rockets on occasion, a tactic more familiar from Afghanistan.

Insurgent mortar teams would often arrive at a location where they had previously dug in a base-plate position, allowing a mortar to be quickly and easily set up. The insurgents would fire a fire round and disappear before either an infantry patrol or helicopter gunships could reach them. Traditionally, the response would have been to employ counterbattery radars to pinpoint the mortar fire and respond with counterbattery artillery fire, a process that U.S. Army artillery can accomplish before the incoming mortar rounds have even landed. In the counterinsurgency environment of Iraq, such a response was unacceptable.

As an example of one such operation, an 82nd Airborne Division sniper team from the 3rd Battalion, 509th Parachute Infantry led by Sergeant First Class Brandon McGuire deployed into an area that had been plagued by insurgent mortar fire. The 82mm mortar team had been firing into populated areas, wounding civilians. The four-man sniper element was dispatched in an attempt to neutralize the mortar threat.

The team covertly inserted overnight into a disused commercial premises that they termed the tractor factory, infiltrating after the curfew to minimize the chance of compromise. Once inside, the team established their covert hide site in a tower within the complex with commanding views over the local area.

After three frustrating days in the hide, their patient wait finally paid off. A suspect individual was spotted digging a cached mortar tube out of the bank of the canal that ran east to west of their hide site. The 40 power spotting scope was employed to confirm the item in the man's hand was indeed an 82mm mortar tube, and permission was gained from their headquarters to engage the insurgent. The team's laser rangefinder displayed a range of 1,300 meters (1,422 yards), certainly within the range of the team's .50BMG M107 but a difficult shot due to both the inaccuracy of the M107 at range against human-sized targets and the high crosswinds that could send the bullet off course.

It took Sergeant McGuire more than an hour to eventually be in a position to take the shot with a good chance of success, as the insurgent kept moving location. Finally he secured a solid sight picture of the man's center of body mass and triggered the M107. The sniper could only see a dust cloud erupt where the insurgent had been standing and, doubting he had hit, he prepared to reengage. His spotter was more certain, however, and confirmed the kill.

DESIGNATED MARKSMEN IN IRAQ

As we have previously mentioned, the Designated Marksman and Squad Designated Marksman programs were largely an American innovation pioneered by both the Army and Marines at roughly the same time. In Iraq, the ranges were not as long as in Afghanistan, but the DM was valued for his ability to provide precision fire in lieu of a sniper assisting in reducing the risk of collateral damage, and for their magnified optics that were often used to identify insurgents hiding among civilians at far greater ranges. It also gave each infantry platoon and eventually each infantry squad a trained countersniper capability.

During the counterinsurgency in Iraq, the need for school-trained snipers far outstripped the numbers graduating from the U.S. Army course, so mobile training teams from the Army's Marksmanship Unit traveled to deploying units to provide tuition on the essentials of long-range shooting and the use of optics and rangefinders. Major John L. Plaster commented: "Though the training is not sniper-qualified, both services provide additional training to its Designated Marksmen, with the Marine Corps course lasting some 23 days. Designated Marksmen have proven invaluable due both to their accurate gunfire and their optical ability to see their surroundings in more detail than ordinary infantrymen."[20]

One of the other advantages of the DM was that in many cases they carried a rifle that visually resembled the standard M16. This assisted in hiding them in plain sight and not attracting the insurgents as a sniper might. Indeed some snipers would follow this example, patrolling into their hide sites with an M4 or M16 in their hands to blend in with the other infantrymen. Of course the DMR also had the advantage of being able to be pressed into service as an assault rifle.

The British did not have a dedicated DM rifle or training program for much of the Iraq War, but used their Light Support Weapon, the L86A2, as a kind of DMR. Although superbly accurate thanks to its longer barrel and bipod, the LSW still fired the 5.56x45mm round, limiting its effectiveness, although anecdotally British marksmen have made hits with the weapon out to a very respectable 800 meters (875 yards).

SOF SNIPERS IN IRAQ

SOF snipers played an important role both during the initial invasion, as we have seen, and later during the insurgency. For instance, the British SAS famously conducted a sniper-initiated ambush against a suicide bomber cell in Baghdad in July 2005. Deploying into a covert hide in civilian dress, the snipers waited for the bombers to leave their safe house. Signals intercepts tipped off the snipers, and, as the three emerged, intent on their mission of murder and mayhem, they were engaged in a coordinated, near-simultaneous shoot that saw all three dead in seconds. The SAS snipers had used the .338 Lapua to devastating effect with headshots on all three, shutting down their nervous systems before they could trigger their devices.

SEAL snipers were also heavily committed to the fight against the insurgency. Along with mentoring Iraqi forces, the SEALs would also deploy on standalone counter-IED and overwatch missions. To camouflage their presence, the SEALs fell back on an old Northern Ireland technique by infiltrating into their target area hidden in plain sight among a mounted infantry patrol in Bradley Infantry Fighting Vehicles. The Bradley carrying the snipers would slow and lower their ramps to let the snipers leap out. On one such mission, the SEALs were compromised within moments of arrival and called back the Bradleys, only for one vehicle to be rammed by a suicide VBIED. Incredibly, none of the SEALs or soldiers were wounded.

SOF snipers became an integral part of the assault forces as they hunted the terrorists of al Qaeda in Iraq. When a target package was developed, the sniper team leaders would be briefed on the location and given access to ISR footage or imagery taken by UAVs or surveillance aircraft. From that, they began to plan where they would locate their teams so that they could provide maximum protection for the assault elements. If the mission used a ground-assault force, the snipers would infiltrate the last kilometer or more on foot to avoid spooking their quarry. They would use collapsible assault ladders to climb to their vantage points. Once the snipers were in position, the assaulters would be given the green light to breach the door.

The SOF snipers were there to do two things: to provide a set of eyes on the objective and to ensure no insurgents escaped as the assaulters cleared the target. Often insurgents would be posted on rooftops as sentries or lookouts and had to be eliminated by the snipers. In a number of operations in the early days of the Iraq War, insurgents had posted grenades down on assault teams from the roofs and upper stories of target buildings. SOF sniper teams, whether they were SEALs, Rangers or British SAS, soon learned that the rooftop was a vulnerability that had to be carefully managed. A Ranger sniper commented: "It was usually a threat from somebody on a roof who could pop up and drop a frag (grenade) down in a courtyard, so I was trying to prevent that and anybody coming off the objective."[21]

A 2008 operation by the Rangers illustrated how their sniper teams operated. Hunting a high-value target, the Ranger assault force drove to the operational area in their M1126 Stryker Infantry Carrier Vehicles.

Halting some distance from the objective to retain the element of surprise, the assaulters crept toward the target on foot. At the same time their four-man sniper team worked their way through the urban maze to reach their previously identified hide site. The snipers used a ladder to access the roof and set up 360-degree security, with two snipers facing the target house.

Two insurgents stationed on the roof heard the breachers preparing their charges on the main door and stood up, AK47s in hand, to investigate. Immediately both were engaged and killed by the Ranger snipers. The assault team successfully breached the house and swept through, killing a number of insurgents, including one who attempted to detonate a suicide bomb vest. The high-value target himself attempted to escape onto the roof but as he emerged he was also shot and killed by the sniper team.

Operations in some notorious areas such as Sadr City or al Qaim would inevitably end in gunfights. One SOF sniper explained that as soon as the shooting started, everyone who had an AK47 would appear on the streets or rooftops hoping to shoot an American. The SOF sniper teams, equipped with their suppressed rifles and night vision technology, would end up killing large numbers of combatants as the assault teams cleared the target building.

According to author Chris Martin, in another operation north of Samarra in 2006, the high-value target was disorientated by the brown out caused by the assault force's Blackhawks as they landed next to his house. He stumbled out carrying an AK47 and was engaged at close range by two Ranger snipers with their SR-25s. According to the snipers interviewed by Martin, the snipers didn't even use their optic, instead aiming by the infrared beams of their lasers only visible through their night vision goggles.

It also wasn't unusual for the Rangers to dispense with the role of spotter, as a Ranger sniper explained to author Gina Cavallaro:

Most people think of sniper teams as shooters and spotters ... The way Rangers operate is, everybody's pretty much a shooter, there's no spotter in an urban environment. You're covering a large area, you're shooting and doing it all

A U.S. Marine Scout Sniper team with 1st Battalion, 4th Marine Regiment, Regimental Combat Team 1 providing overwatch in Al Shur, Iraq during the invasion in 2003. The sniper carries the 7.62x51mm M40A3. Note that both are wearing the MOPP3 NBC suits.

A British Army spotter from 42nd Royal Highland Regiment (Black Watch) in Basra, southern Iraq, 2004, uses his spotting scope to watch for targets for his sniper.

A U.S. Army sniper from 24th Infantry Regiment, 25th Infantry Division provides rooftop surveillance and overwatch in Mosul, Iraq 2004. The sniper uses a tripod mount for his 7.62x51mm M24 mounting the AN/PVS-10 Day/Night Sight. His spotter in the background is armed with a 7.62x51mm M14 with Leupold Mk4 optic.

U.S. Marines in Habbaniyah, Iraq with the 7.62x51mm M40A1 recovered from an Iraqi sniper shot and killed by Scout Snipers from 3rd Battalion, 5th Marine Regiment. The M40A1 was a Marine weapon stolen from an ambushed sniper team in Ramadi in 2004.

The aftermath of a sniper team from 3rd Battalion, 5th Marine Regiment engaging an insurgent sniper team in a civilian car in Habbaniyah, Iraq. The broken glass shows that multiple Marine 7.62x51mm match-grade rounds penetrated, killing both the insurgent sniper and his observer who was videotaping him shooting at Coalition forces. Note the camouflage-painted 7.62x51mm M40A1 the insurgent was using, stolen from a murdered sniper team two years earlier.

A U.S. Air Force Explosive Ordnance Disposal operator in full EOD-8 bomb suit carrying the .50BMG Barrett M82A1 antimateriel rifle during EOD operations in Iraq, 2004.

Although labelled as a Marine Scout Sniper in an overt hide site in An Najaf, Iraq in 2004, the author suspects this is more likely a Navy SEAL working alongside the Marines. Note he has taken off his body armour. The weapon cannot be positively identified, but it appears to be a suppressed 5.56x45mm Mk12 Mod 0 Special Purpose Rifle that was commonly carried by SEAL snipers at the time.

The classic image of a U.S. Marine Scout Sniper in Fallujah, Iraq during Operation *Phantom Fury*. The sniper, from 5th Marine Regiment, 1st Marine Division, is set back in a room to avoid detection and is resting his .50BMG M82A1 antimateriel rifle on a pilfered mattress that will reduce the amount of dust kicked up by the weapon. Notice the ear protection worn against the literally deafening bark of the M82A1.

Another classic sniper image from Fallujah in 2004. This shows a Scout Sniper from 8th Marine Regiment, 1st Marine Division using a loophole to observe for insurgent targets for his 7.62x51mm M40A3. He keeps back in the darkness and doesn't betray his position by poking the barrel out of the loophole.

A U.S. Navy Explosive Ordnance Disposal operator maintains his marksmanship at Ad Diwaniyah, Iraq in 2006 with his .50BMG Mk15 Mod 0 antimateriel rifle.

A Recon Marine Sergeant discovers an insurgent weapons cache in Zaidon, Iraq in 2006. The insurgent rifle appears to be a Mauser of unknown caliber with a crude homemade suppressor and what appears to be a Russian PU scope. Sadly the Marine pictured, Master Sergeant Aaron Torian, was killed by an IED in Afghanistan in February 2014.

A U.S. Army Designated Marksman with his newly-issued 7.62x51mm Mk11 Mod 0 with Leupold Mk4 optic, fold-down backup iron sights and KAC vertical foregrip in Hawijah, Iraq in 2006.

An Australian sniper from 7RAR maintains overwatch from his rooftop hide with his 7.62x51mm SR98 sniper rifle during the SECDET XIII rotation in Baghdad, Iraq 2007.

A U.S. Army Designated Marksman from the 504th Parachute Infantry Regiment, 82nd Airborne Division scans the ridges for insurgents with his Trijicon ACOG-equipped 7.62x51mm M14 during operations in Adi Ghar, Afghanistan 2003. Such M14s were the first true DMRs (designated marksman rifles) deployed by the U.S. Army.

A U.S. Army sniper from the 35th Infantry Regiment, 25th Infantry Division overwatches through the AN/PVS-10 Day/Night Sight mounted on his 7.62x51mm M24 Sniper Weapon System in Afghanistan, 2004.

A Dutch Army sniper team fires a warning shot near a suspected insurgent spotter in Uruzgan Province, Afghanistan 2008. The rifle is the .338 Lapua Magnum Accuracy International Artic Warfare Magnum (AWM).

A well-used 7.62x51mm SR-25
semiautomatic rifle sits ready for
use in a sandbagged bunker at Tarin
Kowt, Afghanistan in 2008.

U.S. Army Designated Marksmen from the 503rd Infantry Regiment, 173rd Airborne armed with 7.62x51mm M14 Enhanced Battle Rifles prepare to engage insurgents in Wardak Province, Afghanistan in 2010.

An Italian Army sniper trains at Combat Outpost Sigma, near Bala Morghab, Afghanistan in 2010. His rifle is the Finnish Sako .338 Lapua Magnum TRG-42 with Schmidt & Bender PMII optic. Note the ejected casing caught in the air as the sniper works the bolt.

A U.S. Army Ranger providing overwatch for a patrol with his suppressed 5.56x45mm SOPMOD Block 2 M4A1 carbine with ELCAN Spectre optic in Zabul province, Afghanistan, 2010.

A Dutch sniper team deployed in the dasht (desert) of southern Afghanistan using camouflage-net Ghillies to break up their shape. The rifle is the .338 Lapua Magnum Accuracy International Artic Warfare Magnum.

An Australian 2 Commando sniper in Uruzgan Province, Afghanistan 2010 armed with a suppressed 7.62x51mm Mk11 Mod 0 in a unique camouflage-pattern finish.

A British Army sniper from 1 Royal Welsh pictured in 2010 in Nad-e-Ali District, Helmand Province with his .338 Lapua Magnum L115A3.

A U.S. Navy SEAL sniper watching for a target somewhere in Zabul Province, Afghanistan, in 2011. His rifle is the SEAL favourite, the .300 Winchester Magnum Mk13 Mod 5 with Mk11 suppressor.

Australian Commandos of the Special Operations Task Group in action in southern Afghanistan. The sniper closest to the camera is firing the .338 Lapua Magnum Blaser R93 Tactical 2 mounted on a tripod, while his colleague is suppressing the insurgents with his 7.62x51mm MAG58 general-purpose machine-gun. Note the spotting scope and M4A5 carbine discarded behind the sniper.

A U.S. Army Designated Marksman aims his venerable 7.62x51mm M14 in Laghman Province, Afghanistan, 2011. The M14 is unusual in its original wooden stock and the lack of a Picatinny rail or optic.

An Australian 2 Commando sniper deployed high on an overwatch position scans for insurgents with his Vector 21 laser rangefinder, his .338 Lapua Magnum Blaser R93 Tactical 2 sniper rifle beside him, somewhere in southern Afghanistan in 2011. Note the range card taped to the stock.

A British Army Platoon Marksman from 1st Battalion, The Princess of Wales Royal Regiment provides overwatch from an FOB in Lashkar Gar, Helmand Province, Afghanistan in 2012. His weapon is the 7.62x51mm L129A1 with 6 power Trijicon optic. The Trijicon is fitted with the co-slaved Shield Firepoint mini red dot sight for close-range shooting.

A U.S. Army sniper providing overwatch security in the mountainous terrain of Kunar Province, eastern Afghanistan, 2012. The sniper's weapon is the recently adopted .300 Winchester Magnum M2010 upgrade of the veteran M24. This image gives some indication of the difficulties in shooting at altitude in such conditions.

Victoria Cross winner SASR sniper Trooper Ben Roberts-Smith with a 7.62x51mm Mk14 Mod 0, fitted with both an EOTech close-combat optic and a swing-out Aimpoint magnifier for longer-range shooting. Note the second sniper in the helicopter behind him with 7.62x51mm SR-25.

A Navy SEAL sniper provides sniper overwatch in Day Kundi Province, Afghanistan 2012. His rifle is a suppressed 7.62x51mm Fabrique Nationale Mk20 Sniper Support Rifle.

A French-manufactured .50BMG PGM Hecate II mounted with a Hensoldt 6 to 24 mildot variable optic and 7.62x51mm American M14 with Schmidt and Bender 3 to 12 variable-power optics used by Estonian snipers in Helmand Province, Afghanistan 2013.

A U.S. Air Force Special Tactics operator with a 7.62x51mm Fabrique Nationale Mk20 Sniper Support Rifle in Khost Province, Afghanistan, 2013. Note the SSR mounts a Bushnell variable-power optic along with an AAC sound suppressor and AN/PEQ-15 infrared laser illuminator.

Afghan National Army snipers under instruction from Scout Snipers of the 3rd Battalion, 7th Marine Regiment in Helmand Province, Afghanistan 2013. The rifle is a suppressed 7.62x51mm Marine issue M40A5.

A Scout Sniper with 1st Battalion, 9th Marine Regiment during operations in Helmand Province, Afghanistan 2013. The sniper carries the 7.62x51mm M40A5: however, note also the three 30-round 5.56x45mm magazines on his plate carrier, indicating he also carries an M4A1 carbine for personal protection.

An Australian Army Designated Marksman chats with village children during a dismounted patrol in Tarin Kowt, Uruzgan Province, Afghanistan 2013. He carries the 7.62x51mm HK417 with 6-power Trijicon ACOG with co-slaved Ruggedized Miniature Reflex (RMR) sight, vertical forward grip and AN/PEQ-15 infrared laser illuminator.

A U.S. Army Ranger sniper firing a suppressed 7.62x51mm Mk11 Mod 0 in Balkh Province, Afghanistan 2014.

An Australian 2 Commando sniper prepares for an operation. His .338 Lapua Magnum Blaser R93 Tactical 2 is strapped to his back with the stock folded, and he carries a 5.56x45mm M4A5 carbine in his hands.

A British Army sniper from the Royal Irish Rangers with his .338 Lapua Magnum L115A3 on board a RAF Merlin. The goggles and face mask help combat the dust and grit blown up by the rotors. Weapons are always aimed downward on aircraft as a safety measure against a negligent discharge.

An Afghan National Army sniper peers through the scope of his 7.62x54mm Romanian PSL sniper rifle, a derivative of the Russian SVD Dragunov.

A Norwegian Army marksman armed with the 5.56x45mm HK416 in Afghanistan. His 416 mounts a Trijicon ACOG, angled vertical grip, folded bipod and what appears to be an infrared laser illuminator unit above the barrel.

U.S. Army Ranger snipers zero their weapons somewhere in Afghanistan. The Ranger closest to the camera has a suppressed 7.62x51mm M110, while the Ranger in the background appears to have a Mk17 SCAR, again equipped with suppressor.

NATO snipers at the Urban Sniper Course conducted by the International Special Training Center in Germany. A range of sniper rifles can be seen. The American .300 Winchester Magnum M2010 and the German .300 Winchester Magnum G22 are in the foreground.

A Green Beret sniper from the U.S. Army's 3rd Special Forces Group trains with his .300 Winchester Magnum Mk13 Mod 7 fitted with a Mk11 suppressor and Schmidt & Bender optic. Note his range or DOPE card taped to the stock.

A sniper instructor from Scout Sniper Platoon of 2nd Battalion, 1st Marine Regiment takes aim at a buoy target with his 7.62x51mm M40A5 sniper rifle during an aerial platform shooting exercise. The M40A5 mounted the Schmidt & Bender PMII variable power optic. Note also the adjustable cheek piece on the stock.

A dramatic shot of a Danish sniper firing his suppressed 7.62x51mm HK417 through a glass window to demonstrate the angle required to minimize the effect on the bullet's flight path, at the International Special Training Center in Germany.

A British Army sniper pair from 1st Battalion, Royal Irish Regiment, displaying their stalking skills during a joint exercise with U.S. forces in Germany. Note the Ghillie suits festooned with local foliage and scrim netting used to break up the outline of the rifle.

An Australian Army sniper, armed with a well-camouflaged 7.62x51mm SR98, prepares to fire from the "laidback" shooting position.

British Army sniper (center) flanked by two French Army snipers during a joint exercise. The British sniper carries a suppressed .338 Lapua Magnum L115A3, whilst the French snipers carry a 7.62x51mm FR-F2 equipped with a Sword Thermal Day/Night sight (left) and a .50BMG PGM Hecate II antimateriel rifle (right).

A SASR sniper rapid-firing his 7.62x51mm HK417 during urban operations training. Note the angled foregrip mounted near the muzzle. This is provides a steadier grip than handguards and is preferred by some over vertical-forward grips.

A U.S. Air Force sniper pair wearing full Ghillie suits while on exercise in Germany. The sniper mans the 7.62x54mm M24 while his spotter is armed with the 7.62x51mm M21.

A close-up image of the .338 Lapua Magnum Blaser R93 Tactical 2 with Schmidt & Bender PMII optic. Note the unusual straight-pull bolt that facilitates fast follow-up shots.

A U.S. Army Ranger provides precision fire to initiate a building assault during training at Fort Bragg, North Carolina in 2012. His weapon is the 7.62x51mm M110.

A U.S. Army Special Forces sniper engaging an explosive target downrange with his .50BMG M107 loaded with the incendiary-tipped Mk211 Raufoss rounds.

A Green Beret from the 7th Special Forces Group zeroes his newly issued .300 Winchester Magnum M2010 with Leupold Mk4 optic.

Marine Scout Snipers line up a shot with a 7.62x51mm M40A5 during the Mountain Scout Sniper Course at Marine Corps Mountain Warfare Training Center in California. Note the specialist digital snow camouflage suits that hark back to similar Russian suits worn during World War II.

ABOVE: U.S. Marine Scout Snipers train with the .50BMG M107A1 antimateriel rifle; the training environment is evident due to the helmets and body armour piled nearby.

RIGHT: The view through a 20-power magnified optic mounted on a 7.62x51mm M40A5 sniper rifle looking at a target 300 yards distant.

U.S. Army snipers from the 501st Infantry Regiment, 4th Infantry Brigade Combat Team receive instruction on delivering aerial sniper fire from a Blackhawk helicopter. The sniper closest to the camera reloads his 7.62x51mm M110, while his partner fires his .50BMG M107 from the open door. Note that both snipers and weapons are tethered to the aircraft.

The cockpit of Air France Flight 8969 at Marseilles Airport, 26 December 1994, in the aftermath of the successful GIGN assault on the hijacked Airbus A300. The majority of the bullet holes in the cockpit glass are from GIGN snipers firing their 7.62x51mm FR-F1 sniper rifles while the smaller holes in the fuselage are from the indiscriminate fire of the terrorists' AK47s. The snipers managed to account for three of the terrorists, who were trapped in the cockpit by the assault team.

GIGN snipers pictured more than two decades after the successful resolution of the Air France 8969 hijacking, this time in January 2015, providing sniper overwatch of the printing works where the two terrorists responsible for the Charlie Hebdo murders holed up in the French village of Dammartin-en-Goele. Hours later the two terrorists emerged in a suicidal charge against the surrounding GIGN operators. The spotter on the right is armed with the 7.62x51mm Heckler and Koch HK417, while the sniper to the left is equipped with the .338 Lapua Magnum Accuracy International AWM.

on your own. There are two snipers with a couple of privates for security ...
The goal is to set up a hidden place before the platoon arrives at the objective.

Another Ranger sniper told her: "It's shooting out to three hundred meters, you have your gas gun [SR-25], it's close range, and you've got to both be shooting on four sides of a building, so you've got a couple of guys for security."[22]

Sometimes the assault force would arrive directly on the target via helicopters. For these missions, snipers would be dropped off by helicopter at a previously plotted location that offered good visibility and lines of sight to the target. On other missions, snipers remained in the helicopters to provide aerial surveillance and precision fire on escaping insurgents. This technique was not without its dangers, and two British Pumas crashed while carrying SAS sniper teams on raids in Iraq, killing a number of aircrew and SAS operators. (Indeed a later coronial inquest took the SAS to task for all but forcing the pilots into dangerous urban flying for which they had received little if any specialist training, unlike many of their American counterparts.)

An example of one such mission in 2005 saw a troop from M Squadron of the British SBS (Special Boat Service) and a U.S. Army Ranger platoon conduct an assault against a terrorist safe house containing suicide bombers. The SBS assault element was met by the first suicide bomber as they advanced on the target, who detonated himself without causing any casualties to the Coalition forces. A second suicide bomber ran from the back of the house and an RAF Puma helicopter swooped in to hover, giving an SBS aerial sniper team on board the chance to kill the suicide bomber before he escaped. The last suicide bomber also attempted to escape on foot but was killed.

The Americans greatly expanded the use of the tactic of AVI or aerial vehicle interdiction. In Iraq this was often conducted by Delta Force snipers who would fire armor-piercing rounds from their 7.62x51mm SR-25s into the engine block of a wanted terrorist while flying alongside the car in a Blackhawk or Little Bird helicopter. If the terrorist didn't stop, the sniper would switch his fire to the driver. Once the vehicle was stopped one way or the other, a second helicopter carrying an assault team would land to detain any survivors and search the vehicle for intelligence materials.

British conventional snipers were also often used to support SOF operations by UK Special Forces, particularly in southern Iraq. In fact, a dedicated infantry platoon was established to do just that. One of the sniper roles that the regular snipers performed for the SOF was a stay-behind surveillance tasking: essentially a sniper team would be covertly infiltrated into an area ahead of a planned SOF raid. Once the raid was completed and the SOF withdrew, the sniper team stayed in place to watch and report upon the effect of the operation.

Sometimes UK Special Forces snipers would repay the favor and lend support to British Army snipers. Sergeant Dan Mills reported one such incident during the defense of the besieged CIMIC House in 2004. Insurgents had learned to stay back beyond the effective range of their 7.62x51mm sniper rifles, and help was requested from Baghdad. The help arrived in the form of a UK Special Forces sniper and a Royal Marine spotter. They brought with them a .50BMG Accuracy International AW50, more than capable of engaging targets beyond 1,000 meters. Soon after, the SBS sniper went to work, eliminating large numbers of insurgents.

At the time of writing, in November 2015, Kurdish Peshmerga forces claimed that U.S. SOF snipers had been heavily involved in combat in northern Iraq. "In February, for the first time, four American snipers came to south Kirkuk because we had lost several Peshmerga to the ISIS snipers. The Peshmerga snipers were weak and before we could hit a single ISIS sniper, we would lose a few men. Therefore we desperately needed American snipers. They had taken part in all the fights in south Kirkuk and they had really good snipers," explained a Peshmerga officer quoted in media reports. Soon after, U.S. Defense

Secretary Ashton Carter announced that what he termed a "specialized expeditionary targeting force," shorthand for a Delta-led Joint Special Operations Task Force, would be deployed to operate into Syria, hunting high-value targets.

TRAINING THE IRAQIS

The pre-invasion Iraqi Army had little sniping tradition within its ranks, as we shall see in the following chapter, instead following the Russian marksman principle of equipping one man in each platoon with an SVD. Training a sniper capability for the New Iraqi Army had to start from scratch, with Coalition advisers conducting modified versions of the basic courses they ran in their own countries.

A U.S. Army sniping instructor explained:

Our snipers trained the Iraqis in all the proper basics of sniper training, so we can eventually take them out on mission and they can start covering their own sectors. We're training them on all the fundamentals – from stalking, shooting, target detection, gathering dope on their weapons and firing at longer ranges than they normally would.[23]

"At first, it was somewhat of a culture shock for the Iraqis, going from their everyday army life to living the life of a sniper," admitted one of the instructors. "They're finally starting to understand that, as sniper, you won't always have ideal conditions, or even ones you'll like. You have to make do with what you have,' he said.

The Iraqis even constructed their own Ghillie suits. They were taught how to camouflage themselves in both rural and urban environments and the fundamentals of the stalk. "We started out with a preparation phase of about 500 meters," explained one trainer. "A sniper can stalk anywhere from 200 to 3,000 meters before settling into their final firing position. From there, they engaged their presented target."

"They're obviously better off now than they were at the beginning," he said. "We gave them a good baseline. However, being a sniper, you have to keep up on your sniper skills. It's very important, due to it being a perishable skill. If you don't practice, it'll fade away." Sadly, based on the performance of Iraqi Army snipers in the fight against Islamic State, much of that training has indeed perished.

CASUALTIES AND COMPROMISES

As Coalition sniper tactics also became better understood by the insurgency, hide sites would regularly be compromised in record time, often by children who were rewarded by the insurgency for exposing the infidel snipers. The Coalition snipers had their own countermeasures. Some areas with a high density of insurgent activity would be saturated with multiple sniper teams in a relatively small area, so that they could both provide mutual support to each other and ensure that if one hide site was compromised, the team could safely extract under their comrades' rifles. After a number of high-profile ambushes of Army and Marine snipers, they also were forced to expand the size of their teams to include security elements.

The first such incident of the Iraq War was in April 2004. A sniper element from the U.S. Army's 1st Cavalry Division narrowly escaped death or capture after they were surrounded and swarmed by insurgents in an apparent organized operation to counter the snipers. In June of that year, in Ramadi, a four-man Marine sniper element attached to Echo Company, 2nd Battalion, 4th Marines weren't so lucky. On an overwatch mission and deployed in a covert hide on a partially constructed building, they were apparently overrun by insurgents and executed.

Three were shot in the head and the fourth had multiple gunshot wounds across his body. One had his throat slit. From examination of their weapons, it was apparent the Marines had been surprised by their attackers and not fired a single round before they were killed. It emerged later that the dead Marines had even been videotaped by the insurgents for propaganda purposes. The team leader, the only Marine among the four to have completed Scout Sniper School, was reported to have told his wife prior to the fateful operations, "they're sending us to the same place, by the same route at the same time of day."

The investigation into the deaths discovered that the insurgents had likely posed as building contractors, perhaps tipped off to the snipers' presence by the building's owners or local sympathizers. "Most likely, about the time the sun came up ... four hit men were probably posing as construction workers to work on the house, and all four were allowed to come up onto the roof, ultimately taking the Marines by surprise," said a source close to the investigation. It was also found that an insurgent cell (mobile) phone conversation had been intercepted by signals intelligence that discussed the impending attack.

The deaths of the Marines were only discovered after a QRF was dispatched due to a failed radio check. Sadly it appeared the QRF wasn't dispatched until some 51 minutes after the team failed to report in for their scheduled SITREP (situation report). The insurgents made off with much of the team's weaponry and equipment, stealing two 7.62x51mm M40 sniper rifles, two 5.56x45mm M16A4 assault rifles, eight fragmentation grenades, a PRC-119F SINCGARS radio and an AN/PAS-13B Thermal Weapon Sight.

Just over a year later, in August 2005, another sniper team was hit, this time in Haditha. A six-man element from the 3rd Battalion, 25th Marines were again executed and again videotaped, although this time the footage was uploaded to the Internet. Two 7.62x51mm M40A5s were also stolen from the dead, along with a number of M16A4 and M16A4s mounting 40mm M203 grenade launchers. The Marines had been operating in two sniper pairs with a third Marine assigned to each team for rear security. The killings were claimed by an insurgent group known as Ansar al-Sunnah.

The Marines immediately added infantry security teams to their sniper elements, although some Scout Snipers argued that this might have been counterproductive as smaller elements could move more quickly and were harder to compromise. Army sniper teams likewise increased security, with a third man being added at the minimum to maintain rear security at a hide site. The environment simply allowed the insurgents too many opportunities to close in on a sniper team unseen.

Along with the size and composition of sniper teams, the rules of engagement in any conflict will always be a source of contention for snipers. In southern Iraq, British snipers experienced ROEs that certainly challenged their skills and patience. Indeed, the insurgents soon began to learn the limits of the ROEs and did their best to exploit them to their advantage. "On many occasions," said a British officer, "Muqtada [al Sadr] militia stood on rooftops from where they had fired in the past, with rocket-propelled grenades and small arms at their feet."[24]

The Iraq War also saw the implementation of a controversial sniper program that was pioneered by the Asymmetric Warfare Group (AWG), a unit of former SOF members who advised regular units on counterinsurgency and counterterrorism tactics. The AWG would provide units with ammunition boxes of "drop items," including objects used in IED construction and detonation that would be placed in areas of significant foot traffic. Any fighting-age males that picked up one of the objects was fair game and could be engaged. A number of snipers from the sniper platoon of the 1st Battalion, 501st Infantry Regiment instead used these "drop items" retroactively, placing them on bodies of Iraqis after they had shot them as justification for the shoots.

Three snipers from the unit were eventually charged with murder; two were eventually convicted of lesser offenses and one found guilty of murder. The AWG had also advised units to dig holes that resembled holes dug by Iraqis to emplace IEDs. If an Iraqi tried to place something in the hole, he could be engaged as likely trying to lay an IED. Such tactics need to be clearly explained in terms of the governing rules of engagement and exactly what constituted hostile intent. At that point in the campaign, the ROE allowed suspected insurgents using mobile (cell) phones or portable radios to be engaged if there was reasonable suspicion that they were attempting to initiate an IED – however, significant doubt surrounded the legality of "drop items" indicating the hostile intent needed to fire.

WARFIGHTING IN FALLUJAH

The Iraqi city of Fallujah had become a veritable insurgent outpost, with foreign fighters lured by al Zarqawi and al Qaeda in Iraq. The insurgents and terrorists effectively controlled the city, but it was not until the murders of four Blackwater contractors in March 2004 that the U.S. military decided to intervene and began offensive operations under Operation *Vigilant Resolve* to recapture the city. Marine Scout Snipers would prove pivotal in protecting Marine infantry on the streets and alleyways of Fallujah.

Marine sniper Corporal, now Sergeant, John Ethan Place was among them. Place notched up 32 kills in the notorious Jolan District during the battle using his 7.62x51mm M40A3 with fixed 10 power Unertl optic. Attached to a Marine infantry company to provide sniper overwatch, he would position himself on the flat roofs of the house in Fallujah, excavating a tiny loophole to shoot through. His longest shot was a headshot on an insurgent at some 515 yards (470 meters).

The young corporal engaged an insurgent car that was speeding toward Marine positions. At 600 yards he fired his first round, striking and killing the driver. The car slowly idled to a stop. An insurgent leapt from the rear of the vehicle but was killed by Place's spotter. The passenger also made a move, running out and firing his AK47 down the street. Place caught him with his second round, dropping the insurgent. According to the sniper, the whole incident took around eight seconds. Place was later awarded the Silver Star, the second highest decoration for valor, for his actions in the city. The sergeant major of the Marine Scout Sniper School, the legendary William S. Skiles, put it simply: "He didn't kill

32 people. He saved numerous lives by protecting our perimeter. That's how the Marines look at it."[25]

During that first Battle of Fallujah, Operation *Vigilant Resolve*, small teams of Delta Force Recce snipers were attached to Marine infantry units to provide expertise and assist in dealing with insurgent snipers. One such team went forward to assist in instructing a Marine unit on a thermobaric AT4 rocket that would later prove deadly against fortified insurgent positions in the city. While operating with the Marine platoon in the Jolan District, their position was discovered by the insurgents, who thought the Americans were the lead element of an attack into the city and consequently poured their fighters toward the two houses occupied by the Marines and Delta operators.

In the subsequent battle, the insurgents virtually surrounded the houses, with only the presence of Delta Recce snipers covering one aspect stopping them from completing the encirclement. The Marine platoon and the seven operators faced up to 300 insurgents, with over half of the Marines wounded in the first deadly minutes of the fight as RPGs rained down on the buildings. Delta Master Sergeant Don Hollenbaugh defended one house solo after his comrade, the late Sergeant Major Larry Boivin, was wounded multiple times and was eventually evacuated from the roof but nonetheless continued firing from a downstairs window. Eventually Hollenbaugh fought back the insurgents, enabling him and the Marines to fall back.

Despite the heroic efforts of the Marines and attached SOF, Iraqi politics and an effective propaganda campaign by the insurgents cut the offensive short with the enemy still in charge. Marine units ringed the city awaiting a change in the political winds. The change arrived in November 2004 when Operation *Phantom Fury* was launched. Both SEAL sniper teams and Army Special Forces were deployed to support the Army and Marine infantry.

Snipers from SEAL Team 3, 5 and 10 were employed as both snipers and to call in close-air support. Chris Kyle was among the SEAL snipers, adding an additional 19 kills to his personal record in Fallujah. He also made his longest shot, with his .300 Winchester Magnum, at an incredible 1,463 meters (1,600 yards). The snipers also benefited from new ROEs that allowed them to engage spotters with mobile phones or radios and those who were laying or triggering IEDs by command-wire or remote-control devices.

A Marine officer described the relationship between the SEALs and the Marines:

We had done the same thing down in Najaf and it worked really well in Fallujah as well. These sniper teams come with a real good capability because they're all Joint Tactical Air Controller [JTAC] qualified and can call fires also. So we put them with our companies and when we got to a key piece of terrain – an extra tall building or a real good field of fire – we'd dismount them and put them in there. We'd use that cover to get them into position and secure them. We'd give them Quick Reaction Force [QRF] capability, so if they had to break contact, they could get back to you.

The style of operation conducted by U.S. Army, U.S. Marine and SOF snipers in Fallujah corresponded well with the manual.

Snipers are attached to friendly units to provide immediate direct support by means of precision rifle fire. The main function of attached snipers will be the suppression of enemy crew-served weapons, enemy snipers, and command and control personnel. Snipers can also support the offensive by interdicting follow-on or reserve forces (such as second echelon combat forces or logistics).

Conventional snipers, assigned to their parent units, can also be used to interdict key targets in the main battle area. Also, attached snipers can be used to screen the flanks of advancing units, cover dead space from supporting crew-served weapons, and engage specific selected targets of the defending enemy units.

During ceasefire negotiations in Fallujah, the insurgents demanded the snipers were withdrawn, a measure of their effectiveness. Along with surgically engaging insurgents, Coalition snipers in Fallujah were also used as forward observers to call in indirect-fire assets such as mortars and artillery.

Sergeant Herbert B. Hancock, a Marine Reservist Scout Sniper who worked as a fulltime Texas SWAT sniper, made the longest known shot with the 7.62x51mm in Fallujah. During the first days of Operation *Phantom Fury*, Iraqi insurgents were flanking Marine resupply convoys. "The insurgents started figuring out what was going on and started hitting us from behind, hitting our supply lines," he said in a 2005 interview with the *Marine Corps Times*.[26]

"Originally we set up near a bridge and the next day we got a call on our radio that our company command post was receiving sniper fire. We worked our way back down the peninsula trying to find the sniper, but on the way down we encountered machine gun fire and what sounded like grenade launchers or mortars from across the river." Hancock and his spotter identified insurgent spotters in civilian vehicles, correcting the fall of shot from the insurgent mortars.

The Marine Sergeant explained:

The insurgents in the vehicles were spotting for the mortar rounds coming from across the river so we were trying to locate their positions to reduce them as well as engage the vehicles ... There were certain vehicles in areas where the mortars would hit. They would show up and then stop and then the mortars would start hitting us and then the vehicles would leave so we figured out that they were spotters. We took out seven of those guys in one day.

Hancock and his sniper were then dispatched to go after the mortar teams themselves.

We moved south some more and linked up with the rear elements of our first platoon. Then we got up on a building and scanned across the river.

We looked out of the spot scope and saw about three to five insurgents manning a 120mm mortar tube. We got the coordinates for their position and set up a fire mission. We decided that when the rounds came in that I would engage them with the sniper rifle.

We got the splash [the first ranging mortar round] and there were two standing up looking right at us. One had a black [outfit] on. I shot and he dropped. Right in front of him another got up on his knees looking to try and find out where we were so I dropped him too. After that, our mortars just hammered the position, so we moved around in on them.

"We adjusted right about fifty yards where there were two other insurgents in a small house on the other side of the position," explained Hancock's spotter, Corporal Geoffrey L. Flowers.

There was some brush between them and the next nearest building about 400 yards south of where they were at and we were about 1,000 yards from them so I guess they thought we could not spot them. Some grunts were nearby with binoculars but they could not see them, plus they are not trained in detailed observation the way we are. We know what to look for such as target indicators and things that are not easy to see.

After we had called in indirect fire and after all the adjustments from our mortars, I got the final 8-digit grid coordinates for the enemy mortar position, looked at our own position using GPS and figured out the distance to the targets we dropped to be 1,050 yards!

Marine Scout Snipers from 3rd Battalion, 1st Marines along with attached SEAL snipers also had a lucky escape. The Marine snipers had established hide sites on the top floor of a six-story building with the SEAL deployed to the roof. After a day of heavy contacts with the enemy, the Marines and SEALs were rotated through to get some rest. Marine Sergeant Matthew Mardan and two other snipers manned the position, regularly checking insurgent approach routes.

Suddenly the snipers were fired at, but this was no insurgent. Instead an Army Bradley Infantry Fighting Vehicle had spotted the three, who were wearing watch caps rather than helmets, assumed they were an insurgent RPG team and opened fire with its 25mm automatic cannon. Luckily the rounds missed, as the snipers frantically took cover, radioing a desperate ceasefire call. Moments later, the wall they had been crouching behind bust open as a TOW II antitank guided missile fired from the Bradley struck the building. Incredibly, the missile failed to explode and the Marines and SEALs raced to safety. EOD later destroyed the missile in a controlled explosion that took the top of the building with it.

THE DEVIL OF RAMADI AND THE LEGEND OF CHRIS KAYLE

As mentioned earlier, Chris Kyle was a Navy SEAL sniper who served multiple deployments to Iraq, eventually notching up at least 160 kills, the most by any U.S. service member (a Royal Marine, whose name is withheld for security reasons, apparently holds the current world record at 173). He left the SEALs in 2009 after four tours of Iraq, including some of the worst fighting in Fallujah and Ramadi. Returning to the U.S.A., Kyle opened a private shooting school. In his downtime he volunteered to help veterans suffering from wartime PTSD by taking them shooting. It was one such veteran who murdered Kyle in 2013.

Kyle is widely divisive, even in death. His most ardent supporters have elevated him to near sainthood, while his most critical detractors argue he was a deeply flawed character with potential psychological issues. With the amount of column inches written about the man, and the differences between his own personal account in his book *American Sniper* and the film of the same name, and the accounts of other SEALs, it's difficult, particularly since Kyle's tragic passing, to separate the myth from the reality. Certainly a number of accounts of incidents that were attributed to Kyle have been widely questioned.

SNIPING IN IRAQ

These included the sniper's assertion that he had shot and killed two men who tried to carjack him back in Texas, and a similar tale that involved him sniping looters and other ne'er do wells in the immediate aftermath of Hurricane Katrina in New Orleans. These may simply have been cases of the sniper mis-remembering details or indeed those with him getting the facts confused. One story that was included in early printings of his book recounted Kyle punching an unnamed former UDT/SEAL celebrity for disrespecting the SEALs at a wake for fallen SEAL Michael Monsoor, who had been posthumously awarded the Medal of Honor for throwing himself upon an insurgent grenade in Iraq to save his comrades. That individual turned out to be wrestler Jesse Ventura, who pursued litigation against Kyle's estate and won.

The "Devil of Ramadi" or "al Shaitan" tag that Kyle said was bestowed upon him by Iraqi insurgents appears also to be difficult to verify independently, as is, surprisingly, Kyle's actual kill tally. The military, including the SEALs, don't keep any sort of official tally from their snipers and actually discourage it, with long memories of the repercussions of body-count reporting from Vietnam.

Despite all of this, Kyle was certainly a very accomplished and courageous sniper who cared deeply for his fellow SEALs and Marines. Many of Kyle's shots were apparently taken without a spotter, which is not unusual for the SEALs, as we have noted earlier, but is nonetheless impressive. Kyle's longest-range shot was 1,920 meters (2,100 yards), against an RPG gunner in Ramadi.

He also accounted for at least one double kill in Ramadi when he engaged a pair of insurgents on a moped who had laid an IED, with the .338 Lapua Magnum round punching through the driver to kill the passenger. Such a shot was rare, but, as we've seen, was accomplished several years later by a British sniper in Helmand using the same caliber of sniper rifle. Kyle was also briefly investigated for that particular shoot as no IED could be found, but a nearby Marine unit vouched for Kyle's version of events.

173

The film version of Kyle's book itself takes significant liberties with the historical record (and with the written source material), although many would argue that this is to be expected from a Hollywood production. It appears to sit uneasily with some, including fellow SEALs, who argue that Kyle's character is placed in situations in the movie that he was never involved in, or worse yet, conducting actions that were actually carried out by other SEALs.

Indeed, even the opening scene in the film, where Kyle's character is forced to engage a child running toward a Marine patrol with an RKG-3 antitank grenade, was apparently manufactured by the film-makers. In his book, it is a woman carrying the grenade who was approaching a patrol during the initial invasion phase near An Nasiriyah that Kyle shoots, his first kill with a borrowed .300 Winchester Magnum Mk13.

The film also claims that when Kyle is deployed to support operations in Fallujah, the alleged bounty placed on his head by local insurgent commanders is attracting foreign fighters who were apparently "flooding the borders to collect on it." Certainly, foreign fighters were entering Iraq in huge numbers in 2004, and many of them were using Fallujah as a base, but they weren't there for Chris Kyle or indeed for any SEAL sniper. We've already mentioned the scene involving the suicide car bomber that wasn't reported by Kyle, but bears a marked similarity to Marine Scout-Sniper Sergeant Tim La Sage's encounter with a VBIED in Ramadi.

The movie did at least portray a number of things historically correctly, such as snipers engaging IED emplacers, the use of apparent civilian spotters by the insurgency to take advantage of the ROE, and it's certainly true that many of the military failings of the film are common to nearly all such productions. The author would argue that, apart from some very nice sniper rifles, the film's only major positive contribution was that it shone a light on military PTSD and the effects of war service on a man and his family.

It was not the first recent film featuring SEALs to manufacture scenes from whole cloth: the previously mentioned *Lone Survivor* inflated the count of insurgents arrayed against the four-man SEAL reconnaissance team and inserted a wholly fictional finale of an insurgent attack against the village that had sheltered the eponymous SEAL, complete with Apache gunships coming to the rescue like latter-day cavalry. In fact, *American Sniper* featured a similarly wholly fictional and faintly ridiculous climactic scene. One wonders what the real Chris Kyle would have made of it all.

CHAPTER 5

SNIPERS OF THE INSURGENCY

The U.S. Army recognizes three distinct types of sniper. One is the school-trained specialist. The second is the trained marksman such as the Designated Marksman. The third type is the armed irregular:

He may have little or no formal military training but may have experience in urban combat. He may or may not wear any distinguishing uniform and may even appear to be merely another of the thousands of noncombatants found in a large urban area. He may or may not carry his weapon openly and may go to great lengths to avoid identification as a sniper. His fires are normally not accurate, and he seldom deliberately targets specific individuals.

His actions are not normally integrated into an overall enemy plan, although his attacks may be loosely coordinated with others in his general area. Although this type of sniper has the least ability to cause heavy losses among U.S. forces, he has high value as an element of harassment, and in some stability and support situations he may achieve results far out of proportion to his actual ability to cause casualties.

The late firearms authority Greg Roberts explained the emergence of the insurgent marksman: "their use of precision fire is in stark contrast with the normal technique of merely leaning around a corner and hammering away with a Kalashnikov in the direction of the 'infidels' in the belief that Allah will guide the bullets."[1] Certainly the great majority of insurgents faced by Coalition forces in war zones like Afghanistan, Iraq, Syria or Mali are thankfully untrained, if perhaps enthusiastic, amateurs. A small proportion will have some form of training, either former regime elements like the Iraqi Army or through some kind of terrorist training camp. Others may have some skill developed from extensive combat experience, for instance Chechen terrorist veterans of the two Chechen wars.

IRAQ

The insurgent or terrorist sniper lacks a number of advantages enjoyed by the military-trained sniper. The Western sniper will have been extensively trained for a minimum of three months in a competitive training environment, training that is constantly updated to take into account real-world operational experience. The Western sniper will also be trained in the use of technological aids like spotting scopes, specialist ammunition and bullet drop compensators not available to the average insurgent gunman.

Insurgents instead took the obvious countermeasure to this lack of training and equipment and simply reduced the range at which they fired. Particularly in the urban environment of Iraq, the insurgent snipers would rarely engage over 200 meters (218 yards) and typically much less than that: if shooting at a target with significant armored cover, such as a HMMWV gunner behind a gun shield, they would often try to shoot from no further than 50 meters (54 yards) away. Obviously any reduction in range tends to increase the chance of a hit. Coupled with even the rudimentary optics of the Soviet PSO scope found on SVD and similar rifles, this enabled the insurgent sniper to become a legitimate threat to Coalition forces.

A Special Forces sniper from the 10th Special Forces Group interviewed by Cavallaro explained:

> Poor snipers don't last very long. The ones that were any good did a pretty good job of engaging targets, and not necessarily from any long ranges, but they did well in concealing their hide sites. For Iraqis it's a lot easier to move among the people. The guy who blends in has much more freedom of movement than we do. It doesn't take that much [either] to make a good hide site, especially in an urban environment if you know what you're doing. They can isolate a target, pick a window, and all they have to do is maneuver into a position so that they can see that one target and shoot him from one hundred meters away.[2]

Iraqi insurgent snipers tended to fire until they registered a hit, even if that round was stopped by vehicle or body armor. At that point, the insurgent would use his local knowledge, and in some cases the active support of the local civilian population, to seemingly disappear into the rabbit warren of back alleys, avoiding further confrontation. Parallel patrolling and similar tactics by Coalition forces sometimes hindered their escape, but ultimately, and in time-honored tradition, the best countermeasure for a sniper was another sniper.

Many Iraqi sniper attacks were also filmed as an integral part of the mission. As mentioned earlier, insurgent sniping is as much about propaganda as it is about engaging and killing the enemy. Video cameras became as much of an insurgent signifier as an AK47 or RPG. Even if the sniper was unsuccessful in killing a Coalition solider, the video of the incident could be cleverly edited and posted to the Internet or distributed on wildly popular DVDs that were available in Baghdad street markets. An insurgent on one such video even explained their thinking: "The idea of filming the [result of] operations is very important because the scene that shows the falling soldier when hit has more impact on the enemy than any other weapon."

In one of the most disturbing attacks of the war in August 2007, a U.S. Army unit from the 1st Battalion, 30th Infantry of the 2nd Brigade Combat Team was fired upon by an insurgent marksman. The shooting appeared to have been a ruse to draw the soldiers into the building where the marksman fired from. Once inside, a large HBIED (House-Borne IED) consisting of four looted Iraqi Army 155mm artillery shells was was detonated, killing four soldiers. Graffiti in the ruins taunted the survivors with "This is where the sniper got you guys!" Such attacks were planned with one eye firmly on their propaganda value.

We've mentioned a number of myths and fallacies that surround sniping, particularly when it is portrayed in Hollywood films. A recent culprit has been *The Hurt Locker*, which is interesting as it features insurgent sniping in one key scene where a group of British military contractors and American EOD operators are engaged by an insurgent sniper team.

The scenario shown in the film is nonsensical at best. The Coalition forces inexplicably cannot initially locate the source of the insurgent sniper fire, despite there being only a single building in the otherwise largely featureless desert that could be harboring the insurgents, and despite their vastly superior optics and training. The British team even have a .50BMG Barrett M82A1 ready to engage any long-range targets, but the canny insurgent sniper soon makes short work of the British sniper.

In fact, the insurgent sniper is making shots with apparent ease, killing three of the contractors before they can start to effectively respond (although the Iraqi sniper had evidently never heard of the concept of eye relief, as he fires with the scope of his FPK within about an inch of his eye: being Hollywood there is luckily no recoil to give him a nasty bruise). The Americans eventually take over the rifle, with one of the EOD soldiers acting as spotter calling the range as "850 meters" (930 yards), well within the range of the Barrett but a tough shot with the FPK the insurgents seem to be using, even by a school-trained sniper.

The EOD soldiers go on to eliminate the Iraqi snipers and their security team with a string of shots from the Barrett, including dropping a running insurgent at that same range of 850 meters (930 yards), which would make a Marine Scout Sniper proud let alone an EOD operator. Contrast the apparently competition-grade insurgent sniper team featured in the film with the accounts of actual Iraqi snipers, and you will soon see the massive gulf between fiction and reality.

Iraqi insurgents did find inspiration in strange places, however, and followed in the footsteps of both the Provisional IRA and the Washington D.C. Beltway Snipers when they began using a sedan as a covert sniping platform. These mobile sniping platforms became common enough for the U.S. Army to document the tactics of the insurgent *Portable Shooter* in a 2012 briefing:

1. Surveillance will confirm static location and choke point.
2. Range to target will be measured or estimated.
3. Driver positions vehicle.
4. Video cell may record the shooting.
5. Shooter confirms specific target from camouflaged platform viewpoint and shoots.
6. Vehicle moves calmly into traffic pattern and departs area.

In response, veteran sniping authority Major John L. Plaster argued: "I urged the widespread distribution of optics among deployed units to better detect sniper vehicles before they fired, and a countertactic of not waiting for a reaction force, but immediately and aggressively rushing any sniper who fired at American forces, whether with four men or 400."[3]

As noted, the car provided perfect cover for the escape of the sniper team, particularly in heavily populated urban areas. Famously, this didn't stop Specialist Stephen Tschiderer, whose shooting by an insurgent portable shooter was videotaped by the insurgents. Tschiderer was standing next to an up-armored HMMWV in Baghdad providing security when he was shot by an insurgent sniper secreted in a nearby van. The round struck the SAPI trauma plate in his body armor and failed to penetrate. Tschiderer quickly got back up, took cover behind the HMMWV and directed the vehicle's turret gunner toward the source of the fire.

The insurgent-made video is amusing in its incompetence, as the insurgents continue to film despite it becoming increasingly obvious that the American patrol know their location. In fact, the sniper van doesn't move off until a pair of responding HMMWVs are almost on top of them. Moments later, they are captured by the Americans. Hardly the master snipers of *The Hurt Locker*. Intriguingly, the van had been equipped with a number of mattresses in an attempt to deaden the sound of the sniper's SVD being fired.

In other incidents, a Mazda sedan was extensively modified to allow the sniper to fire through a small hole in the trunk in an almost exact copy of the setup used by the infamous PIRA .50 cal sniper. The driver acted as the sniper's spotter due to the dramatically reduced field of vision such a position offered, as well as videotaping the shootings for distribution as propaganda. Another concealed a sniper under rolled carpets in the back of a van with a loophole to shoot from. The insurgents regularly carried a number of stolen license plates that they used to confuse the Iraqi police.

A U.S. Marine sniper team engaged one such portable shooter with surprising results in June 2006. A sniper pair from Sniper Section 4 of the 3rd Battalion, 5th Marine Regiment was providing sniper overwatch as a patrol of Marine AAV-P7/A1 tracked vehicles drove past. The sniper team spotted an Iraqi sitting in the passenger seat of a parked car videotaping the Marine convoy. The spotter then sighted the wooden stock of a rifle. He alerted the Marine patrol and the sniper dialed in the range. With a single shot from his 7.62x51mm M40A3, he killed the insurgent with a headshot.

The insurgent's partner – his spotter, as it turns out – returned to the car. "We then saw another military-aged male ... enter the passenger side door," explained the Marine spotter. "He was surprised to see the other shooter was killed."[4] The insurgent climbed into the driver's seat, but before he could start the car, the Marine spotter engaged him three times with his 5.56x45mm M4 carbine, gravely wounding him, and he died soon after. The nearby Marine patrol searched the vehicle and to their surprise recovered a Marine-issue 7.62x51mm M40A1 sniper rifle. "When we saw the scope and stock, we knew what it was." The M40A1 had been one of the weapons stolen from the 2nd Battalion, 4th Marine Regiment Marine sniper team ambushed and killed two years earlier in Ramadi (as detailed in the previous chapter).

Certainly the insurgency understood the value of sniping. It inflicted a huge psychological cost on Coalition forces, who understandably feared being shot from long range every time they left their FOBs on patrol. They also inflicted undeniable casualties: in one three-year period following the invasion of Iraqi, over 40 American servicemen were killed by insurgent sniper fire. According to a 2005 instruction that appeared on a number of jihadist websites and was translated from the original Arabic by the U.S. Army, insurgent snipers should:

1. Target enemy snipers and surveillance teams.
2. Target commanders, officers and pilots; that is, target the head of the snake and then handicap the command of the enemy.
3. Assist teams of mujahideen infantry with suppressive fire. These teams may include RPG brigades or surveillance teams.
4. Target U.S. Special Forces, they are very stupid because they have a "Rambo complex" thinking that they are the best in the world. Don't be arrogant like them.
5. Engage specialty targets like communications officers to prevent calls for reinforcements. Likewise, tank crews, artillery crews, engineers, doctors, and chaplains should be fair targets.
   - A tank driver was shot while crossing a bridge, resulting in the tank rolling off the bridge and killing the rest of the crew.
   - Killing doctors and chaplains is suggested as a means of psychological warfare.

6. Take care when targeting one or two U.S. soldiers or agents on a roadside. A team of American snipers [may be] waiting for you. They [may be] waiting for you to kill one of those agents and then they will know your location and they will kill you.
7. In the event of urban warfare, work from high areas and assist infantry with surrounding the enemy, attacking target instruments and lines of sight on large enemy vehicles, and directing mortar and rocket fire to front-line enemy positions.

By 2008, the U.S. Department of Defense was asking for more funds to purchase both active and passive sniper countermeasures for Iraq.

The dangers from enemy sniper attacks have increased steadily during the past year, with the number of attacks quadrupling. These attacks have not only caused numerous casualties, but have had an adverse psychological effect on both Coalition forces and the Iraqi civilian populace. Victims in sniper incidents have a fatality rate of over 70 percent. A shift in enemy tactics that increases the number of sniper attacks could potentially inflict even more casualties than IEDs.

Thankfully, this increase in sniping attacks was effectively countered by the massive influx of U.S. forces into Iraq as part of the "surge." With more troops on the ground, and ever more patrols, coupled with the growing Anbar Awakening that saw former insurgents switching sides in preparation for power-sharing, the opportunities for sniper attacks in fact dwindled, although sadly never fully disappeared.

JUBA

In Iraq, insurgent sniping attracted its own myths, and none was greater than the legend of Juba. Juba was a supposed sniper working for the insurgent Islamic Army of Iraq, predominantly in southern Baghdad. A skilled shot, Juba would aim for the areas of the body not protected by the body armor's trauma plates or ballistic helmets. If Juba existed, his reign of terror seems to have peaked between 2005 and 2007.

Despite his oft-reported skill, most of the shoots attributed to him were at 200 meters (218 yards) or less, again hardly the sign of an experienced marksman, although the dense urban terrain of much of Baghdad would admittedly reduce lines of sight to his target, forcing any such sniper to close with his target. Rumors abounded of the death card Juba apparently left at the scene of his kills – "What has been taken in blood cannot be regained except by blood – The Baghdad Sniper." A curiously American affectation for an Iraqi: perhaps he had simply watched too many Vietnam movies?

U.S. troops held widely varying views on Juba at the time. Some believed the claims, while others pointed to apparently equally mythical Chechen snipers who were the preferred bogeymen of the time both in Iraq and Afghanistan. An Iraqi interpreter working for Coalition forces interviewed by *Stars & Stripes* newspaper in 2007 claimed: "He's killed many people. He was trained during the time of Saddam. He can shoot me when I'm walking; he can shoot me when I'm running; he can shoot me in a vehicle. He is very good."[5] The killing of the Marine sniper element in Ramadi in June 2004 was also attributed to Juba by many within the insurgency.

In comparison to the numbers of Coalition soldiers and Marines killed or maimed by IEDs, however, Juba's contribution, and indeed the overall contribution of insurgent snipers, is far less. Juba's value to the insurgency, like the Chechen snipers, was more psychological than physical. Juba also turned the tables on the Americans after the reign of terror their snipers inflicted on the insurgency during operations in Fallujah, for example, where the "men with the long rifles" were rightly feared.

The most likely explanation for Juba is that he is a composite of a number of insurgent snipers of varying quality who operated during the same time period. Supporting this theory were the insurgent videos themselves, which sometimes showed the always masked Juba firing twice rather than his more disciplined single shot before moving. A Green Beret commented: "I've heard the name Juba thrown around; it could be like William Shakespeare though. He could have been real, but did he write all the plays attributed to him?"[6] Pre-invasion, the Iraqi Army trained snipers, of a sort. These were modeled along Russian doctrinal lines, with a marksman with an SVD or similar platform in every squad to provide longer-range capabilities. The U.S. Army noted that "most of these 'trained' snipers are equivalent in skill to a Squad Designated Marksman."

Juba's alleged tally of kills is likewise uncertain: one insurgent propaganda video claimed an astounding 143 deaths of Coalition soldiers at Juba's hands. Those numbers would surely have been enough to have elevated him to the top of the targeting charts for an SOF strike team, yet none of the special operations units committed to destroying the insurgent and terrorist networks in Iraq at the time have ever mentioned a target like Juba. Despite this, Juba's legacy continues to this day, with an Islamic State "sniper battalion" claiming him as an inspirational figure.

The late Chris Kyle made reference to another supposedly legendary insurgent sniper who was elevated to his on-screen nemesis in the film of the same name: "From the reports we heard, Mustafa was an Olympics marksman who was using his skills against Americans and Iraqi police and soldiers," was the description given by Kyle.[7] He claimed that an Iraqi insurgent sniper matching this profile was later killed by other U.S. snipers.

In the film, it is a Syrian sniper, and he should have been easy to locate thanks to the black eye he would have been sporting. The actor in the film again (like in *The Hurt* Locker) has his weapon's scope pressed virtually against his eye socket, with predictable results should it have been firing anything heavier than blanks.

The insurgent sniper that Dillard Johnson had killed in Salman Pak at the range of 850 meters (929 yards) was apparently a Syrian who "was believed to be the same sniper who had killed upwards of 20 American soldiers with some damned good shooting."[8] Johnson concedes that the insurgent was a better shot than he was. Some of the Syrian's shots had been recorded on video and uploaded to the Internet in the manner of Juba. Whether the Syrian sniper was Juba or simply a particularly skilled insurgent, who can say?

Even the British had their Juba, a mercenary sniper known in southern Iraq as the Albanian. "He would shoot out of a vehicle and drive away," explained one sniper. "He was so good that he hit people through the side of their Osprey. He never hit anywhere else, only through the side. We heard there was someone in Bosnia with the same signature."[9] Although this story has never been verified, rumors exist of another trained sniper who was targeted in a countersniper effort by Black Team snipers from SEAL Team 6 who eventually stalked and killed the man, firing the lethal shot through the loophole the insurgent was using.

The U.S. Army saw little evidence, however, for any great number of highly skilled mercenary insurgent snipers, noting a report from August 2004 "of a 'sniper' in Najaf firing more than 80 rounds over the course of 8 hours at U.S. forces, but this sniper's firing did not result in any casualties," while admitting that "the incorporation of scopes on weapons has probably increased the average insurgent's marksmanship out to perhaps 200 to 300 meters. There is evidence of some true snipers operating in some insurgent groups, which is exhibited by spikes in single shots to the head and torso [shots through the side of the IBA]. A possible source of these true snipers might be the influx of experienced veterans from the Iran–Iraq war."

IRAQI SNIPER WEAPONS

Along with export SVDs of Russian manufacture, Iraqi insurgent snipers used the 7.62x54mm al Kadissiya, a locally produced SVD copy, the 7.62x54mm Romanian manufactured FPK and the 7.62x39mm Tabuk, both of these latter two being essentially heavy-barreled AK variants based on the RPK light machine gun but fitted with a Soviet-style

PSO optic. At least one, presumably captured, 5.56x45mm weapon was also being used in southern Iraq in 2007. According to press reports, seven British servicemen fell victim to an insurgent sniper using the 5.56x45mm firing American-manufactured ammunition.

The availability of 7.62x54mm ammunition varied widely and included the 7N1 152-grain match sniper load, the standard 7N13 steel core penetrator and the 7B–Z–3, a 165-grain armor-piercing incendiary. Former Iraqi Army snipers were issued with only 40 rounds of ammunition, five of which were the armor-piercing incendiary, but thankfully there are no confirmed accounts of this being used against the body armor of Coalition soldiers. Indeed, from reports it appeared few if any insurgent snipers understood the different types of ammunition that may have been available to them.

Although it was unproven, the Coalition accused Iran of providing the fearsome .50BMG Steyr Mannlicher HS antimateriel rifle to Shia insurgents in the south who were being covertly trained and supported by Iranian al Quds special operators. In another curious aside, Sergeant Dillard Johnson recovered an iron-sighted .303 MkIII Lee Enfield from an Iraqi sniper during the initial invasion and used this "as a sniping weapon – it worked well for targets within a few hundred meters." He went on to say, "later in the war I was shooting RPG guys at 200 plus meters with it. When that .303 bullet hit them they stayed down for good."[10]

CHECHEN INSURGENTS

According to a RAND Corporation study of the conflict, Chechen snipers in the First Chechen War were called "ghosts" by their Russian adversaries and were much feared for their ability to operate at night. In contrast, a lack of training and batteries for night vision devices, along with poor maps and imagery, precluded the few Russian snipers from conducting similar operations. Consequently, the Chechen "ghosts" owned the night and limited Russian freedom of movement.

The Chechen insurgents benefited, at least during the First Chechen War in 1994, from a large number of insurgents who had served in the Russian armed forces, thereby providing an already trained core for their force. Some of these men were undoubtedly trained as snipers, or in the Russian system as marksmen. As we've noted previously, Russian infantry platoons have long featured a squad marksman who today would be analogous to the Western idea of the Designated Marksman. They tend to receive little to no specialist training, but are trained and equipped with the SVD.

Having received at least some basic military training gave them an advantage over their latter-day counterparts in Iraq, for example. Chechen snipers would use hides in ruined buildings and, in particular, loopholes in attic spaces that would offer maximum concealment, positioned to watch a potential Russian line of approach. Chechens would also use the so-called "come-on ambush" to lure Russian forces into the firing line of a previously emplaced sniper. They would target officers and radio operators as priority targets.

During rural operations, Chechen insurgent snipers increased their engagement ranges up to a reported 1,000 meters with the increased lines of sight available outside the cities. When operating in the rural environment, with less concealment, the Chechens deployed security elements to overwatch their snipers. These would typically feature a four-man element armed with AKs who would engage any Russians who got too close to their sniper, drawing fire and giving the sniper the chance to break contact and escape.

Chechen insurgent snipers also formed a vital element of Chechen tactics against Russian armor. Snipers using SVDs featured as part of their tank-hunter teams. According to Lester Grau, a typical *Panzerknacker*-style tank-hunting team would be composed of three insurgents: one carrying the PKM, another with the RPG and the third with an SVD. The RPG would be used to disable the armored vehicle (often these were fired in concentrated volleys by a number of tank hunters working in concert), while the SVD and PKM were used to discourage any supporting infantry. The SVD would also come in handy to dispatch any tank commander foolish enough to operate unbuttoned in the urban maze of Grozny.

In Fallujah, some insurgent sniper pairs were encountered, although, lacking training, they would invariably both operate as individual shooters rather than the traditional sniper and observer structure. In at least one incident, an insurgent sniper deliberately wounded a Marine crossing an open, vulnerable point, luring others to attempt a rescue, who were themselves then shot in an eerie parallel to the Stanley Kubrick film *Full Metal Jacket*. Similar tactics have been seen with Chechen insurgent snipers. Researcher Olga Oliker notes, "A common sniper ploy was to shoot individual soldiers in the legs. When others tried to help the wounded soldiers, they too came under fire."

White Widows were the alleged Chechen women snipers, the equivalent of the Chechen Black Widow suicide bombers who avenge the deaths of their husbands and boyfriends at the hands of Russian forces. Documented incidents involving Black Widows are few and far between, although a number of Black Widows wearing suicide vests were claimed to have been involved in the infamous 2002 Dubrovka theater siege in Moscow. The Black Widows, part of a Chechen gang of up to 40 terrorists, took over the popular theater while the musical *Nord-Ost* was playing, seizing 912 hostages in the process. Three days later, a Russian Special Forces assault killed all of the Chechens, while a reported 130 hostages were controversially killed by an unidentified incapacitating chemical agent pumped into the building to render the hostage takers unconscious. Confirmed cases involving a White Widow are even harder to uncover.

In 2014, a British tabloid ran the story of one alleged White Widow, although not of Chechen origin: a British terrorism suspect named Samantha Lewthwaite who had apparently been killed fighting as a sniper in the Ukraine. The newspaper based its dubious claims on a Russian media report that Lewthwaite, the wife of one of the London bombing suspects, had become a member of a Ukrainian volunteer battalion fighting against pro-Russian separatists. Further stretching the credibility of the report, Lewthwaite had been apparently dispatched by a Russian sniper in an *Enemy at the Gates*-style showdown. In all probability, the terrorist remains in Somalia with the al Qaeda-linked al Shabaab, to whom she fled in the aftermath of the 7/7 bombings, although other reports indicate she may have joined Islamic State in Syria.

The White Stockings are yet another group of female insurgent snipers who, according to both the Russian security services and tabloids, fought in at least both the First Chechen War in 1994 and the Georgian War in 2008. Olga Oliker explains that these shadowy figures were allegedly "female snipers from the Baltic states, Ukraine, Azerbaijan, and Russia itself, who hire themselves out to the rebels." They apparently used a range of VSS and SVD rifles and would taunt the Russians over the radio, as most Russian military communications during the Chechen wars were, unbelievably, not encrypted and thus could be monitored by the Chechens with simple commercial scanners. The author doubts the reality of either the White Stockings or the White Widows, but they have both served as effective propaganda tools.

CHECHEN INSURGENT WEAPONS

Chechen units used Russian-issue small arms, including their sniper rifles. They were organized into combat groups that somewhat approximated a Western squad, including a PKM gunner (the Soviet 7.62x54mm medium machine gun, roughly similar to the M240 or MAG58), a pair of RPG gunners, three riflemen with AK74s and a sniper with an SVD of which the Chechens apparently had some number, thanks to the withdrawing Russians kindly leaving behind a reported 533 examples. A number of integrally suppressed 9x39mm Vintorez VSS carbines were also later employed by the Chechens in the Second Chechen War. Persistent rumors continue to circulate about their use of commercial .22LR bolt-action rifles with homemade suppressors and subsonic ammunition, although the range and lethality of such a combination raise questions as to the truth of such stories.

AFGHAN SNIPERS

While Afghan insurgent snipers were much rarer than many accounts credit, genuine marksmen did exist and were sometimes used to devastating effect by the Taliban and other anti-Coalition militias. One example lies in accounts of the battle of the Uzbin Valley. A joint force of French, Afghan and U.S. Army Special Forces entered the town of Spur Kunday in the Uzbin Valley in August 2008 searching for a rumored force of foreign Taliban from Pakistan.

A platoon of French paratroopers securing the high ground were themselves ambushed by a much larger insurgent force. Accurate fire from what a later NATO investigation identified as an SVD Dragunov was responsible for the deaths of a number of French soldiers. Tellingly, the fire targeted leaders, radio operators and interpreters in the manner of a school-trained Western sniper.

Or consider the two British paratroopers who were also killed by an insurgent sniper, apparently by one bullet. Both were members of a joint patrol targeting insurgents in the district of Nad Ali in 2011. The insurgent sniper narrowly missed with his first round. As the British troops took what little cover was available, Private Lewis Hendry was shot in the head. The round exited and struck Private Conrad Lewis, who was directly behind him, in the neck. Tragically, both later died from their wounds.

Make no mistake: Afghan insurgent snipers did and do exist and they have certainly killed Coalition forces. More common, however, are the accounts of harassing fire by insurgent snipers. For instance, another apparent sniper appeared in August 2007 in Kajaki Province.

An insurgent marksman had been firing single shots at British patrols. He hadn't hit anyone, but was getting closer and closer by the day. Despite deploying a number of their own snipers against him and conducting a number of reactive OPs (sniper ambushes by any other name), the insurgent was never caught or killed, leading to speculation as to his origins.

Colonel Richard Kemp commented upon the insurgent's fieldcraft in his account of the Royal Anglians' 2007 tour, *Attack State Red*: "The sniper appeared at different points on the battlefield, regardless of the location of the main Taliban forces. They had found some of his skillfully concealed firing positions. He used tunnels to slip from one to another, and his escape routes were impossible to spot from the air." Kemp concluded: "They were up against an excellent sniper, clearly well trained, perhaps by a regular army: their speculation included Iran or Pakistan, and even the possibility that he was a veteran of the mujahideen campaign against the Soviets."[11]

Afghan snipers, or at least Afghan insurgents using long-range rifles like the Dragunov SVD, were typically employed as part of Taliban ambushes, with the SVD joining the PKM and RPG as the weapon of choice for suppressing Coalition forces. The advantage of the SVD, even if inexpertly fired, is of course the ability to engage targets at ranges greater than the standard AK47. As we've discussed earlier, the Taliban and other Afghan insurgents soon learned to increase the stand-off distance between them and the predominantly 5.56x45mm weapons carried by ISAF infantry patrols.

Like Juba, many column inches have also been devoted in tabloid newspapers to the supposedly veteran marksmen among the Afghan mujahideen and later the Taliban, raised with a rifle in hand and able to pick off imperialists of any stripe at astounding ranges. Although many believed the archetype of the grizzled Afghan marksman, invariably armed with some variant of the venerable bolt-action .303 Lee Enfield, there is little actual evidence to support the idea.

New York Times journalist and author of the definitive history of the Kalashnikov series, C. J. Chivers, examined the data around gunshot wounds to American troops, including Marines deployed to Helmand Province, looking for evidence of enemy snipers. He found that gunshot wound lethality in Afghanistan hovered between 12 and 15 percent, far lower than the one in three of America's previous wars, and hardly indicative of a large cadre of well-trained insurgent marksmen.

Chivers handily collected the main problems facing the concept of the mujahideen marksman: "limited Taliban knowledge of marksmanship fundamentals, a frequent reliance on automatic fire from assault rifles, the poor condition of many of those rifles, old and mismatched ammunition that is also in poor condition, widespread eye problems and uncorrected vision, and the difficulties faced by a scattered force in organizing quality training."[12] The Afghans also share something of the religiously inspired fatalism of other Islamic insurgents and terrorists. They will hit the target if God wills it: it is a predestined outcome that such concepts as cheek weld, eye relief or even using the sights will not alter.

We should focus on a couple of the excellent points Chivers identified, however, namely the "old and mismatched ammunition" and the "widespread eye problems and uncorrected vision." Let's tackle the ammunition question first. As an insurgent organization, the Taliban and other anti-Coalition militias in Afghanistan cannot rely on a consistent source for ammunition resupply like a regular army can. Instead, they must scavenge ammunition from wherever they can. While the majority still comes on the back of a mule across the border with Pakistan, some is captured or purloined from Afghan security forces, some is from elderly stocks dating back to the Soviet Afghan War, and still more was supplied by various parties during the Civil War of the 1990s.

The quality of this ammunition consequently differs wildly. Consider also the environment in much of Afghanistan, which is anathema to corrosive ammunition like much of that from the former Warsaw

Pact, along with the exceptionally poor storage and maintenance of insurgent ammunition stocks. All of these factors will detrimentally impact on the accuracy of any insurgent snipers. Compare this with the care afforded to Western match-grade sniper ammunition.

The eyesight argument is also a major factor in assessing the viability of Afghan snipers. The country is still one of the poorest on the planet, and consequently the level of healthcare availability, especially for those outside the major population centers, is essentially nonexistent. Even within urban centers like Kabul, much of what little healthcare is available is provided by charity organizations. Rural healthcare has been a key tenet of the Coalition's counterinsurgency strategy, with reconstruction teams offering clinics staffed by military medics as part of the effort to win the proverbial hearts and minds.

With virtually no preventative eyecare available, poor diet, lack of vitamins, and an increased chance of cataracts from long-term ultraviolet light exposure, Afghans generally suffer from poor uncorrected vision. Such eyesight does not bode well for a sniper. Combined with poor ammunition quality and often badly maintained weapons (the author has seen numerous SVDs recovered from insurgent weapons caches without their optics), the myth of every Afghan male being a sharpshooter is indeed a myth. A U.S. Marine Corps captain interviewed by Chivers summed up his experience of being on the receiving end of Afghan shooting: "I used to say in Iraq that I'm only alive because Iraqis are such bad shots," he said. "And now I'll say it in Afghanistan. I'm only alive because the Afghans are also such bad shots."

As we have seen, insurgent snipers and the stories around them often take on a life of their own. In Sangin, southern Afghanistan, such a mythology developed around a sniper or snipers known simply enough as the "Sangin Sniper." Dependent on the source, the "Sangin Sniper" was responsible for anywhere between five and ten casualties to British and U.S. Marine forces. There does appear to have been an insurgent marksman active in the area in August 2010: at least two of the incidents attributed to the sniper showed some skill-at-arms and the probable use of a magnified optic.

Stories soon circulated that the "Sangin Sniper" was a Chechen mercenary or team of mercenaries from Chechnya, Pakistan and Egypt. A British Ministry of Defence spokesman at the time commented: "What the guys are seeing on the ground, not just in Sangin but also in Nad-e Ali, is an increased use of the tactic of single shots at range. It would be vastly overplaying the professionalism and the effect of these people to call them snipers. But, of course, we're adjusting our tactics to counter it."[13] Tabloid accounts speak of UK Special Forces being deployed in a number of countersniper missions that allegedly resulted in the deaths of the foreign fighters from an airstrike guided in by the Special Forces operators. Like many such stories, independent verification is impossible. Based on some of the reported details however, some of these tales smack more of journalistic invention than accounts of actual operations.

AFGHAN INSURGENT WEAPONS

The most common sniper rifle employed by Afghan insurgents is, not surprisingly, the 7.62x54mm Dragunov SVD. Chinese copies have also surfaced, as have a range of former Warsaw Pact variants. Bolt-action 7.62x54mm Mosin Nagants and .303 Lee Enfields have also been used by insurgent snipers, but they are the exception rather than the rule despite popular culture.

Few advanced sniper weapons have been seen or recovered in Afghanistan, and most of what is available to the insurgents is decades old. One rather exotic weapons system has been seen in propaganda videos, however – the World War II-era Soviet 14.5mm PTRD bolt-action antitank rifle. The PTRD has only been encountered when the insurgents are carrying out a large, preplanned attack on a Coalition base. One was seen being employed in the attack on the combat outpost at Wanat, for instance.

ISLAMIC STATE SNIPERS

Of course no survey of insurgent snipers would be complete without mention of Islamic State, who apparently make significant use of snipers. Numerous videos exist allegedly showing the exploits of Islamic State snipers fighting against the Syrian Army and other insurgent groups in Syria. They have also been extensively deployed in the defense of key cities the terrorists have seized in Iraq.

They routinely parade their so-called sniper battalions who, according to Islamic State propaganda, have been inspired by Juba. Dressed in tan fatigues and plate carriers with a set of locally produced leather magazine pouches for their SVDs sewn onto the carrier, they appear ill at ease with their weapons in comparison with trained Western snipers, whose rifles all but become an extension of their bodies. However, they are still likely effective against other insurgents and most Iraqi military.

During the retaking of Sinjar in Iraq in late 2015 by the Kurdish Peshmerga, supported and mentored by U.S. Army Green Berets, the advance was often slowed by a combination of Islamic State snipers and IEDs. The insurgents would often leave a number of snipers behind to harass Peshmerga forces as they entered villages and towns that Islamic State had withdrawn from, covering the withdrawal and giving their colleagues time to get away. Like the use of Taliban snipers and IEDs in the village of Marjah that pinned down a Marine advance for hours in 2010, similar tactics are now being used in Iraq.

Islamic State Weapons

Islamic State snipers have been seen carrying an unusual American weapon: the 7.62x51mm Mk14 Mod 0 Enhanced Battle Rifle, the modernized version of the classic M14. The Mk14 was provided to Iraqi Army SOF by American advisers and a number have now been captured by Islamic State. Most of their sniper rifles are less surprising: ancient 7.62x54mm Mosin Nagants, SVDs and Chinese-manufactured Type 79 Dragunov copies. The odd captured 5.56x45mm M16A4 with ACOG or commercial optic has also been seen.

In terms of antimateriel rifles, the most common are the Russian 12.7x108mm semiautomatic OSV-96 (or OBC-96) and the Chinese semiautomatic M99 12.7x108mm. The insurgents have also used several homemade platforms, including an apparent 23mm bolt-action rifle in Kobane. Equally fascinating was an SVD or Type 79, mounted in a crude but operational remote weapons station. A number of Croatian 7.62x51mm Elmech EM-992 bolt-action rifles with unidentified scopes have also been employed by Islamic State. These are very traditional, wooden-stocked rifles that would appear better suited for civilian hunting than combat.

CHAPTER 6

POLICE AND COUNTERTERRORISM SNIPING

Although this chapter divides the police and counterterrorist sniper as two different skillsets, they are often one and the same. In most Western nations, for instance, the primacy for domestic counterterrorism response still lies with the police, so law enforcement snipers may well find themselves responding to a terrorist incident, particularly in the current age of the so-called active shooter scenario as witnessed in Mumbai and Paris.

Conversely, counterterrorist snipers may be drawn from either a police or a military organization. There is of course significant overlap between sniper operations conducted during a military counterterrorism campaign in Iraq, for instance, and one conducted in a domestic setting such as a siege or hostage-taking by terrorists. What we are discussing in the following pages is primarily focused on the latter. Law enforcement or counterterrorist snipers also operate in a very different way to military snipers deployed in conventional warfighting or even counterinsurgency.

Firstly, they tend to operate under far more restrictive rules of engagement, or "use of force guidelines," as they tend to be termed by police. They cannot simply engage a target because he or she is armed or is clearly involved in a criminal act. Most police snipers can only fire if they are in fear of their own life or the lives of their fellow officers or civilians. In extreme situations, a police or counterterrorist sniper may be allowed to engage a suspect when they are not imminently threatening someone's life, particularly if the target is holding hostages or is wearing a suicide bomb vest or similar explosive device. In general, however, "police are authorized to use lethal force, but only if their life or the life of another is in imminent threat of death or grievous bodily harm," as a tactical officer and firearms instructor explained to the author.

An instructor at the U.S. Army's Special Forces Sniper Course argued:

We're there to create confusion and interdict targets with precision fire and take them out. By comparison, a police sniper is there to end some sort of situation that's taking place. He's there to put an end to it if he gets the opportunity to shoot, to deliver precision fire to end some type of hostage or crisis situation.[1]

An Australian police tactical officer explained it to the author:

Those split second decisions will then be examined and cross-examined for months by the coroner post incident. This is significantly different from the military. Again operations [domestically] are completely different legislatively, politically and from a community expectation point of view. This will be on the mind of all members in any incident, in that making sure their decision is the right decision and that they can justify and explain it in coroner's court after the event.

Secondly, the police or counterterrorist sniper will normally engage their targets at far closer range than their military counterparts. Of course there have been and will be exceptions to this; however, the majority of the shots taken by police snipers will be within an urban environment that limits engagement ranges. Retired American SWAT officer Craig Roberts estimated that most police sniper shoots in the United States are around 60 to 70 meters (65 to 75 yards) in range. In hostage situations, the U.S. Army's Special Forces Sniper Course cautions against attempting a headshot beyond 200 meters anyway, as the chances of a hit on the *fatal T* we discussed earlier is low.

Another Australian police tactical officer similarly noted:

Everyone tends to think about the need to shoot accurately over long distances, however, that's not how it works in real life. Most police situations are resolved within 200 meters. It's very rare that a perimeter would be more than 200 meters away, except perhaps in a country location. With police marksmanship, we're not training people to shoot apples at 1,000 meters, we're training them to shoot realistically up to about 300 meters.

Police snipers also have to be far more concerned with and aware of any chance of overpenetration of their target, as they will likely be shooting in heavily populated built-up areas. They must understand the implications of shooting through glass and intermediate barriers such as cars or building walls. Such considerations lead to unique requirements for the type of weapons and ammunition deployed by police and counterterrorist snipers.

Indeed, former British Army sniper instructor Mark Spicer argues that police snipers need a range of platforms in 5.56x45mm, 7.62x51mm, .338 Lapua and .50BMG: "Without them the snipers will find themselves making do with either over-powered or under-powered rifles, with the risk of injury to hostages or the general public, and possible consequential criminal proceedings against them."[2]

In most tactical units, to remain as a police sniper, officers must be able to hit within a specified grouping with their first shot from a cold bore at 100 meters. A cold bore simply means a weapon that has not been previously fired: a weapon that has been fired is warmer and the round will be more accurate and shoot a little further. Most police and counterterrorist sniper shots will be from a cold bore.

Cold bore or not, one of the great myths of police snipers is the oft-touted option of shooting the gun from a criminal's hand. This has occurred in reality, but it is incredibly rare and an extremely dangerous shot to take. Consider what happens if the sniper misses and kills the individual, or the round misses and goes on to harm a noncombatant. At the time of writing, only four such shots have been made by SWAT snipers in the United States, as far as the author can ascertain, and, tellingly, it appears all involved shooting the weapon from the hand of a suicidal person rather than a hostage taker or gunman.

The idea that snipers can simply shoot guns out of people's hands, held by many otherwise sane and sensible individuals, is often accompanied by a stated preference for police to shoot criminals in the leg or arm, a so-called non-lethal option. However, shooting someone in an extremity can be just as dangerous as a solid hit to the center of body mass if an artery is struck. It is also far more dangerous to any hostages, civilians or other police, as a round to an extremity does not guarantee incapacitation of the target. Remember that police snipers should not be firing at all unless they fear for their own lives or the lives of others: they are literally the final option.

The police sniper should be attempting to incapacitate the hostile individual: he or she is not necessarily intending to kill, although this is often the predictable result of high-velocity rifle rounds striking vital organs within the human body. Shots to the center of body mass, as we have seen with military snipers, are both the safest for bystanders and offer the greatest chance of a hit that will take the suspect out of the fight, which after all is the reason the sniper is employed in the first place. An armed individual shot in the arm or leg is still very capable of killing a hostage or bystander. Shots to the center of body mass, or, if tactically conceivable, to the head, are the only way to ensure incapacitation.

The London Metropolitan Police Force Firearms Unit, SCO19, then known as SO19, beat strong international competition using both their 7.62x51mm Accuracy International rifles and their G3Ks to win the 2005 British Army International Sniper Symposium at the Infantry Battle School in Brecon, Wales. It was even more remarkable as many of the scenarios were related to military operational experiences in Iraq and Afghanistan. One of the police shooters explained:

> We faced a number of scenarios, one of which was baling out of a Warrior armored fighting vehicle, with an observer keeping watch from the turret and guiding the sniper to a terrorist target 530 meters away.
>
> Another scenario was based on operating in a built-up area and we were in a fortified house in a sand-bagged position. We could see more than 140 "friendly targets" and in a very short space of time, five of these became hostile targets. It was an exercise in observation and the ability to engage the target accurately and quickly once establishing it was hostile.

SWAT ORIGIN

Although police departments around the world have traditionally employed officers as marksmen, whether suitably trained or not, law-enforcement sniping as we know it today really only began to emerge as a response to the turbulent events of the late 1960s and early 1970s. In the United States, homegrown radical militant groups like the Symbionese Liberation Army (SLA), the Black Panthers and the Minutemen were arming themselves with semiautomatic rifles and shotguns, outgunning patrol officers. Internationally, the scourge of terrorist hijackings and hostage takings also demanded a specialist response. Additionally, police departments were forced to face the phenomena of the mentally unstable gunman, such as Charles Whitman, a former U.S. Marine who shot and killed 14 people with rifle fire from the tower of the University of Texas in 1966.

Domestically, police tactical units were developed to tackle the wave of armed revolutionaries, the most famous being the American

SWAT or Special Weapons and Tactics teams. These units were initially equipped with military fragmentation vests or flak jackets, AR-15 rifles and standard issue pump-action shotguns. The role of marksman or sniper was often simply assigned to the best shot, who was typically armed with either a bolt-action civilian hunting rifle or a variant of the U.S. Marine Corps M40 or M40A1, itself built from the civilian Remington 700.

SWAT's role was to contain barricaded offenders until such time as negotiators could agree a surrender or, if the offenders began harming hostages or firing on police, to conduct a forced entry on the location in an attempt to arrest the criminals. One of the most famous early such incidents was the pitched battle that occurred between the fledgling Los Angeles Police Department SWAT team and SLA militants on East 54th Street in Los Angeles in 1974.

After attempting a call-out using a megaphone, the SWAT officers, including a sniper armed with a civilian bolt-action .243 rifle, came under fire from the inhabitants of an SLA safehouse. The 19 deployed SWAT officers, some using only the curb for cover, returned fire on the house and a prolonged gunfight developed over the course of several hours. Officers fired CS gas grenades (tear gas) into the safehouse. Two SLA members made a suicidal break from the house with revolvers in their hands and were gunned down by the police. The others refused to surrender and the notoriously pyrotechnic CS gas canisters eventually caused a fire that engulfed the house, burning to death three other militants. A fourth apparently committed suicide.

In the decades since, tactical policing has evolved considerably. The best of today's tactical teams, and snipers, are well armed and equipped and have access to first-rate law enforcement and civilian shooting schools. Indeed, many tactical teams are trained to a similar level as military counterterrorist units. Although they primarily deal with criminal acts, with active shooter incidents on the rise thanks to lone wolf terrorism, such skills will be sadly tested in the years to come.

COUNTERTERRORISM ORIGIN

In the United Kingdom and Europe, police forces also raised tactical units that would respond to both criminal and terrorist sieges and hostage takings. In some counties, the counterterrorism response remained with the police, while in others military SOF were given the task. This growth was largely driven by the events in Munich in 1972. In September of that year, Palestinian terrorists from the Black September organization took a number of Israeli athletes and coaches hostage during the Summer Olympics.

After negotiations stalled, plainclothes West German police armed with Walther submachine guns approached the apartments where the hostages were being held in preparation for a covert entry through ventilation shafts. None of the policemen had any specialist training or equipment and their approach was captured on live television, which was seen by the terrorists, forcing the police to retreat. A second, desperate plan was hatched.

After deceiving the terrorists into believing they had acceded to their demands, the German police deployed five policemen, armed with 7.62x51mm Heckler and Koch G3 battle rifles devoid of optics, as snipers, although none had received any training whatsoever. They also had no night vision devices, even though the plan called for the terrorists to be engaged at night. As the terrorists approached a pair of UH-1 helicopters that would ferry them and their hostages away, one of the policemen opened fire prematurely. In the confused gunfight that followed, two terrorists were killed, as was a policeman.

After the terrorists shot out the floodlights illuminating the scene, the police were effectively blinded. Following an uneasy stand-off, one of the terrorists pitched a fragmentation grenade into one of the waiting helicopters, killing three hostages. Another terrorist opened fire on the other helicopter with his AK47, murdering five more hostages. Ultimately, all of the hostages died during the bungled rescue attempt, and five of the eight terrorists were killed. Germany immediately pledged that such a tragedy could never be allowed to occur again. With their military constitutionally barred from operating domestically, the Germans instead turned to their Federal Border Police and raised a new unit, Grenzschutzgruppe 9, better known as GSG9.

From 1973, GSG9 pioneered counterterrorism tactics that were adopted and refined by other units that maintained a close relationship with the Germans including France's GIGN (Groupe d'Intervention de la Gendarmerie Nationale or National Gendarmerie Intervention Group), the Spanish National Police's GOE (Grupos de Operaciones Especiales), the Italian NOCS (Nucleo Operativo Centrale di Sicurezza) and the British SAS. The unit's first major operation, the successful storming of a hijacked Lufthansa airliner in Mogadishu, Somalia, is still studied today. GSG9 is divided into three core operational components based around the insertion method. GSG9/1 is the ground-based intervention force, while GSG9/2 deals with maritime targets and infiltrations and GSG9/3 are trained in parachuting and heliborne assaults.

In response to the growing terrorist threat and in direct response to the United States hosting the 1984 Olympic Games, the Federal Bureau of Investigation (FBI) raised their own counterterrorist tactical unit in 1984, which was christened the Hostage Rescue Team or HRT. Along with a sizable assault element, the 50-man HRT also included a number of what were officially termed countersnipers. In its original incarnation, the HRT was divided into two subunits in a similar fashion to the British SAS who had been involved in their initial training. The Gold and Blue teams each included two seven-man assault elements and an eight-man sniper element.

Even at the elite level of the HRT, police sniping remains a game of patience. Former HRT sniper Christopher Whitcomb explained that "sniping is a lonely, intrusive business built on long hours of boredom and rare moments of epiphany and thrill." He felt that snipers were very much the less glamorous end of tactical policing: "snipers had to lie out in the weeds for days at a time, enduring subzero temperatures or sweltering jungle heat, hoping for an occasional glimpse of the bad guys."

Like many American police units, the HRT went with the bolt-action 7.62x51mm Remington 700 with a 10 power Unertl optic with bullet-drop compensators as their primary sniping rifle in both standard and suppressed configurations. Unlike many military snipers, they are permitted to use match hollowpoint ammunition rather than ball ammunition. HRT snipers were also equipped with 5.56x45mm CAR-15s with Aimpoint optics for close-range work, 7.62x51mm M14s that were used for shooting from helicopters and boats and the .50BMG Barrett M82A1 for extreme distance and vehicle interdictions.

By way of comparison, France's elite GIGN was also formed as a reaction to the events in Munich and to an event a little closer to home, the seizing of the Saudi Arabian embassy in Paris by Palestinian terrorists in 1973. The unit is trained and equipped to conduct armed interventions against terrorists and hostage takers both within France and internationally. Sniping is very much part of the unit's culture. Intriguingly, all members of GIGN complete their basic sniper course with certification in both 7.62x51mm and .338 Lapua platforms. Specialist snipers are further trained on the .50 platforms and attend the full sniper school. GIGN have operated both within France and internationally. We will look at several of their more famous operations later in this chapter.

TYPES OF OPERATION

The U.S. Army divides the tasks of the military CT sniper into three primary functions.

1. They can deliver discriminate fire to interdict hostile targets.

2. They can cover the entry teams into the objective area with rifle fire.

3. They can provide the CT force commander with his most accurate target intelligence. In this case, snipers are normally positioned to have ideal observation of the enemy. Most frequently, this will be the commander's only view of the target.

The police sniper may also carry out all three of these roles, but they may have further tasks such as VIP protection and overwatch of political or public events that the military CT sniper will seldom face. This serves as both a reassurance for the general public and as a high-profile deterrent to any but the most committed terrorists. Often an overt sniper position at such an event will be supported by a covert position nearby.

HOSTAGE RESCUE AND ACTIVE SHOOTERS

The most likely call-out for a police sniper to encounter is hostage taking, generally the result of either a failed criminal enterprise like an armed robbery, or a domestic violence scenario where partners and children may be seized against their will. Police tactical units will deploy an inner cordon to isolate the hostage taker(s) and an outer cordon to prevent friends, family or compatriots reaching the suspects. Snipers can be part of both cordons. Counterterrorism units will be employed at a hostage taking, however, if the situations they are called upon to resolve are politically or religiously inspired.

In both routine barricaded suspect and terrorist incidents, police snipers are first and foremost employed as the eyes and ears of the tactical police. They can often provide an indispensable running commentary on exactly what is occurring inside a barricaded building. This is particularly valuable in the early hours of a siege or similar incident, as technical surveillance equipment will be yet to be deployed. Police snipers are typically veteran members of their tactical unit, as this gives them greater insight into the types of intelligence that will help both the assaulters and the negotiators. Along with intelligence gathering, they are primarily deployed to physically contain the situation and isolate the suspects.

Former tactical officer Mathew Coombes explained to the author that full-time police snipers often conduct all of the covert surveillance and reconnaissance work for their units.

There was limited work for a pure marksman so the team began to do a lot more surveillance and information gathering when building up to a job. This has been a real boon for them and they are now busier than ever.

The marksman can deploy for both rural and urban taskings. This includes the full suite of rural equipment, camouflage and OP skill sets you would expect of a battalion [sniper] team or similar. They can also do this for a number of days. Insertion is generally by vehicle at night and on foot from there. They may go into a location the night before or up to a week before a resolution is planned to help build the "intelligence" picture prior to the resolution happening.

They will have the usual night vision equipment, some thermal and high-resolution optics for this work. In an urban environment they have civilian vehicles, camera equipment and the like to assist in surveillance and information gathering for the lead up to a job. This is quite an effective role as it saves the teams having to do it all and also the need to call in other areas to assist with this, which can be risky given they do not always appreciate or understand the information needed in a tactical job.

The snipers will provide an overwatch function on both entry and exit points from the location should a suspect attempt to flee. Again this is most useful during the early stages of an incident, as assault or arrest teams will be moved into place to cover any openings as soon as is practicable. The snipers will also provide overwatch during the approach of any assault element preparing to breach into a location either as part of a deliberate action or an immediate action plan. During an assault on a stronghold, snipers are often cleared to engage targets of opportunity that present themselves.

Coombes explained:

In your typical siege operation, marksmen will be positioned in an overwatch location. They will also provide commentary in conjunction with the tactical team cordon. Every time they saw the offender, they would have been giving commentary as well. Yes, they will provide overwatch for the assault, but not fire support from the purely military perspective. On this, aside from their sniper rifle, marksmen may have a 40mm launcher.

Now this is obviously not for HE [high explosive], but for gas/Ferret rounds [Ferrets contain CS gas and are designed to be shot through doors and other obstacles: they are also very effective at deflating vehicle tires]. Prior to any planned assault they may be required to fire gas through the windows or other openings of the stronghold just prior to the assault taking place.

Although often portrayed in films, the command to fire should only come from a sniper team leader, not the on-scene commander, hostage negotiator or anyone else in the chain of command. The incident commander, whether police or military, will keep his snipers from firing until a) the situation has deteriorated to the point that the suspects or terrorists must be killed to save lives, or b) a preplanned deliberate action assault is under way and the snipers have been tasked to support it.

Known by a number of names, an Immediate Action or IA plan is one that is enacted in response to the actions of the hostage takers, for instance if they begin harming hostages. Such actions will lead to an immediate assault on the hostage takers' strongpoint. The French Police BRI assault at the Bataclan Theater in Paris in November 2015 or the assault on the Radisson hotel in Mali are recent examples of where the assault was forced by the execution of hostages.

Indeed, due to the rash of active shooter scenarios in the United States and the terrorist incidents in Mumbai and Paris, the protocols for managing such incidents have changed. No longer are responding police advised to emplace a cordon and await a tactical unit. Instead they are encouraged to make entry, preferably using a four-man diamond formation, and attempt to confront the active shooter(s) themselves to save innocent life. This is obviously very dangerous for the officers involved, who have little specialist training or equipment. It also means that the incident will hopefully be satisfactorily resolved with minimal loss of life before the sniper teams even arrive.

In contrast to such immediate action operations, the Deliberate Action or DA is a plan that is developed as intelligence is gained on the location (including such things as the direction doors open, type of locks, and width of walls, for example), the suspects (their demeanor and psychological state, their objectives, weapons and similar), and the physical locations of both the hostage takers and the hostages. The sniper teams can assist greatly with providing this intelligence to facilitate a DA. They may also form an integral part of the plan themselves.

The DA is carried out to a timeline decided by the tactical unit, normally once the negotiators feel there is little chance for a peaceful resolution. Indeed, the negotiators may well assist the DA by tying up at least one of the hostage takers on the telephone while the assaulters move into place. This was famously seen during Operation *Nimrod*, the British SAS assault at the Iranian embassy in London in May 1980. There, the terrorist leader was kept on the telephone talking to negotiators, which stopped him from investigating the strange noises he was reporting. The strange noises were in reality members of B Squadron of the SAS preparing to rappel down from the roof.

As an aside, both police and SAS snipers were deployed at the Iranian embassy, although no sniper shots were taken. It was reported that the initial D11 rifle officers were replaced by six SAS sniper teams, who were ready to either engage targets of opportunity as they presented themselves or to deliver CS gas via grenade launchers. This secondary task is still a common one for both police and CT snipers, as they will often have an ideal vantage point from which to launch either CS gas or flashbangs from 40mm launchers.

The surveillance function performed by the snipers on the cordon will sometimes initiate an entry by a tactical unit. For example, during the infamous Sydney Lindt Café siege in 2014, a sniper witnessed a hostage being murdered by the deranged gunman, triggering an immediate breach by the Tactical Operations Unit that ended in the deaths of both the gunman and a second hostage. As an aside, the same sniper, armed with a .338 Lapua Accuracy International AWM, could not engage the suspect as he would have had to fire through two panes of reinforced safety glass. The glass could have easily impacted his point of aim and placed hostages at risk from his fire.

In police shootings, snipers are generally very reluctant to shoot through glass if there are any civilians or hostages in the vicinity. A shot must strike as close to 90 degrees and as level as possible to the glass to minimize impact on the point of aim. Police snipers will only consider such a shot in a dire emergency, for instance if a hostage is in impending danger of being executed. Even then, if the hostage taker is more than a meter (3 feet) away from the glass, the chance of deflection from a headshot is so great that many snipers would refuse to take the shot as they could easily end up inadvertently killing a hostage.

At distances of up to 2 meters from the glass (6 feet), snipers are reluctant to take even center-of-mass shots in close proximity to hostages. As the distance from the glass increases, so does the margin of error from the intended point of aim. Police snipers also have to consider spalling from the glass that could cause serious wounds to a hostage. Indeed, some snipers argue that the only way to even remotely safely shoot through glass is for a near-simultaneous double shot from two snipers: one to break the glass above the intended target, and thus minimize the chance of a wound to a noncombatant, while the second sniper engages the actual target through the breaking glass. This kind of shooting requires exceptional skill and timing, and could only be attempted by really top-tier sniper teams who have drilled extensively in the technique.

A real-life example of when shooting through glass can go dramatically wrong was the infamous Good Guys electronics store hostage taking in

Sacramento County, California in 1991. A SWAT team sniper, cleared to fire on any armed individual within his line of sight, engaged a gunman tethered to a hostage, but the sniper's round, a standard .308 match-grade round, broke up when it hit the glass of a closing door.

A waiting SWAT entry team was to assault into the store upon the sniper's shot (a sniper-initiated entry). Inside the building, four Vietnamese gang members held staff and customers hostage. The gang began shooting hostages and had killed three by the time the entry team managed to conduct a breach and shoot three of the suspects dead, wounding a fourth. Ammunition has since been developed to try to minimize the effects of glass on the trajectory of a sniper bullet. Bonded (where the lead core has been soldered to the bullet's outer jacket) or solid brass bullets assist with minimizing the dispersion caused by shooting through glass, but there is no perfect solution.

One of GIGN's first missions involved the unit's snipers firing through glass – specifically automobile glass – in Djibouti, East Africa in February 1976. At the time, Djibouti was known as the French Territory of the Afars and the Issas, named after its two largest ethnic groups, and was still administered as a French territory. A terrorist organization known as the Front for the Liberation of the Coast of Somalia (FLCS) was fighting for independence from France.

On February 3, 1976, a school bus carrying 30 French children, dependents of locally based French Air Force personnel, was seized by four FLCS terrorists carrying AK47s. The terrorists drove the bus to the border with Somalia, picking up a fifth terrorist along the way. When they attempted to cross the border, however, they were stopped by a unit from the French Foreign Legion based in Djibouti. The terrorists' demands were farcical: they wanted the independence of the territory from French control, otherwise they would begin executing the children. An uneasy stand-off ensued.

The operation was complicated by its location, right across the land border with Somalia. In fact, Somali border guards were actively assisting the hostage takers. Indeed, within a short distance from and overlooking the hijacked bus was a Somali border-control post and a sand-bagged machine gun position mounting a fearsome relic of World War II, a 7.92x75mm MG42 capable of firing up to 1,200 rounds a minute. The legionnaires were forced to aim their weapons at both the Somali border guards and the FLCS terrorists.

GIGN were immediately dispatched from France to effect the hostage rescue of the children. Their plan was simple but audacious. As part of the negotiations, the children would be supplied with sandwiches containing a sedative that would send them to sleep, effectively removing them from the line of fire. Once the sedatives took effect, a team of GIGN snipers would simultaneously engage the five terrorists with headshots, while a GIGN assault force ran to the bus and rescued the children. The legionnaires would keep the Somalis at bay with suppressive fire from their AA52 machine guns and 60mm mortars.

At just before 16:00 local time, with the children dozing, the snipers struck. Near simultaneously, the five snipers fired their 7.62x51mm FR F1s, ensuring that the angle of the shots was as close as possible to a perfect 90 degrees to reduce the risk of the glass windows sending either glass spalling or parts of the bullets into the young hostages. Five terrorists fell dead, all shot in the head. While the GIGN assaulters dashed for the bus and the legionnaires took the Somalis under suppressive fire, a solitary Somali border guard managed to get close enough to the bus to fire a burst into the vehicle, killing a five-year-old hostage before he was gunned down by GIGN. A sad end to an otherwise exemplary sniper-led hostage rescue.

Along with the dangers of shooting through glass, the police or CT sniper also obviously needs a clear and unobstructed line of sight. After the January 9, 2015 Paris siege at the Vincennes kosher supermarket, members of the French National Police RAID (Research Assistance Intervention Dissuasion) tactical unit explained why their snipers could not be used to engage the AK47-armed terrorist inside: "our snipers could not risk taking a shot at him, because there were advertising posters all over the windows and it was impossible to get a clean shot."

Additionally, the terrorist, Amedy Coulibaly, had lowered the aluminum security shutters that covered the majority of the windows, making a clear line of sight almost impossible. Eventually an assault team from RAID and the BRI breached into the supermarket, and, amid a rain of flashbang grenades, shot and killed the terrorist before he could murder any further hostages (he had killed four earlier in the siege). Coulibaly ran at the RAID officers and was shot some 40 times at the entrance to the supermarket, collapsing at their feet.

At the same time, some 45 minutes from Vincennes in a printing plant in the village of Dammartin-en-Goele, the two French Algerian terrorists responsible for the Charlie Hebdo murders of January 7, 2015, Saïd and Chérif Kouachi, were holed up, surrounded by France's elite Gendarme counterterrorist unit, the GIGN. The Kouachi brothers, armed with AK47s and wearing German Army-issue combat body armor, were seemingly preparing for a final shootout with police. GIGN snipers armed with .338 Lapua Accuracy International AWM rifles and 7.62x51mm HK417s manned rifle posts on surrounding buildings.

The pair evidently chose "suicide by cop" and exited the factory, wildly firing their AK47s. They were engaged by both the GIGN inner perimeter team and multiple snipers, who killed both brothers in a fusillade of shots. Unsure if the terrorists had further collaborators, GIGN launched an immediate assault into the factory and rescued the single hostage, who had been hiding from the terrorists throughout the incident. Two GIGN operators were lightly wounded during the firefight. GIGN EOD operators reportedly found a live grenade positioned as a final booby trap under the body of one of the brothers.

In France there is a bewildering number of specialist intervention units with counterterrorist, hostage-rescue and anti-kidnapping duties. The BRI (Brigades de Recherché et d'Intervention or Anti Gangs

Brigade) for instance, is a Paris-based unit tasked with confronting organized crime, including criminal kidnappings. The BRI can assist in providing operators trained in basic hostage rescue techniques and did so in support of RAID, the French national police's SWAT or tactical unit which is deployed to resolve both terrorist and criminal sieges.

In the case of the Vincennes and Dammartin-en-Goele sieges where significant manpower was required at geographically dispersed locations (along with maintaining tactical support for the investigators into the Charlie Hebdo murders and elements to deal with any further contingencies that might have occurred in France or overseas), both RAID and BRI were deployed along with GIGN to resolve the incidents. Similar raids to arrest terrorists usually feature a sniper element to provide overwatch and surveillance for the assaulters. In another recent example from France following the events of November 2015 in Paris, elements from RAID conducted an explosive entry on an apartment in Saint Denis, a northern suburb of Paris. Inside were a number of armed terrorists with AK47s and suicide-bomb vests. Unfortunately, the explosive breaching charge did not topple the reinforced door and it was only partially breached.

Having lost the element of surprise, the RAID assaulters were immediately engaged by the terrorists and were forced to pull back, while the terrorists rolled an apparently homemade ballistic bunker into place behind the remnants of the door. Previously unknown to the French authorities, a second apartment was being used by the terrorists and they began receiving a fierce crossfire. The gun battle swiftly escalated, with the terrorists firing hundreds of rounds from their AK47s and lobbing fragmentation grenades into the hallway. Five RAID assaulters were wounded. Six snipers had been deployed on adjacent buildings providing sniper overwatch and intelligence to the entry team. After a lull in the firing, the snipers reported little movement.

According to the commander of RAID, the unit then sent one of their combat assault dogs forward with a camera mounted on his harness to conduct a reconnaissance. Sadly the terrorists shot and killed the dog. Moments later, one of the terrorists appeared at an apartment window firing his AK47. He was engaged and shot by a RAID sniper, who inflicted a wound that would kill him sometime later from the resulting blood loss. Another terrorist then appeared at a window firing his AK47. After a pause, and in an apparent attempt to destroy the building and bring it down upon the heads of the RAID officers, he detonated the suicide bomb vest he was wearing.

While the blast did not topple the building, it significantly weakened the structure and blew holes in the floor. Grotesquely, part of the suicide bomber and a female accomplice who also perished in the blast were blown from the window and landed on a police vehicle in the street below. After firing a number of 40mm flashbang grenades into the apartment, a micro UAV (drone) was introduced. It is believed the terrorist shot by the sniper had already expired by this point. Further reconnaissance by EOD robots was blocked by rubble from the terrorists' grenades, and the suicide bomb blast forced RAID to use pole cameras (infrared capable video cameras mounted on extendable poles) to survey the apartment from below through the holes in the floor.

Eventually the objective was secured and two further terrorists hiding in the apartment complex were arrested. The RAID operation highlighted the use of police snipers in a very close-range barricaded suspect operation. The snipers acted as the assaulter's eyes and ears, as they had much better visibility from their rooftop hides. Judging by bullet strikes on the wall of the apartment building, it appeared also that a .50BMG antimateriel rifle, most likely the French PGM Hecate II, was employed to try to shoot through the wall to hit terrorists on the other side.

Russian counterterrorist tactics are somewhat more brutal than those practiced by Western units like GIGN, where the overriding emphasis is on the safe release of hostages and, if possible, the suspects themselves. To some observers, the Russians often seem far more concerned with killing all of the terrorists, or "bandits," as they refer to them. Along with the infamous Moscow theater siege, a terrifying example is the June 1995 seizure of a hospital and a number of municipal buildings in the Russian border town of Budennovsk by approximately 200 Chechen terrorists.

The Russian SOF operation to recapture the hospital was straightforward, but showed less concern for the hostages they were attempting to rescue. While MVD Spetsnaz units conducted suppressive fire against the front of the hospital, the Alfa hostage rescue team attempted to conduct a breach during the confusion. They recaptured a portion of the building and killed a number of Chechen snipers, but then became pinned down by the terrorists. A number of hours later a similar assault was launched, with sadly similar results. Incredibly, however, through a negotiated settlement, the Chechens were allowed to withdraw from the town unhindered. The civilian death toll was estimated at 150.

The Russian Alfa unit has also employed snipers as the primary tool to end a hostage taking. In July 2001, a passenger bus carrying 40 civilians was hijacked by a Chechen gunman, demanding the release of a number of Chechen prisoners held in Russian jails. Along with an AK47, the terrorist was carrying hand grenades and a significant quantity of dynamite. Negotiations released a number of hostages, but the terrorist grew increasingly agitated, firing his weapon in the air. A hostage rescue was ordered.

Alfa deployed a number of snipers to cover the careful stalk of the members of the assault element, who crawled carefully into position in the tall grass along the side of the road. Once the assaulters were in position and night was falling, two flashbang grenades were detonated on the road in front of the bus. The hijacker rushed forward, giving the snipers a clear shot, and he was engaged by multiple snipers. Hit four times, including once in the head, the terrorist fell forward onto a live grenade he had been carrying as it exploded. Apart from minor injuries from flying glass, the hostages were released unharmed.

VIP AND CLOSE PROTECTION

VIP close protection has some similarity to military sniper overwatch, but is often conducted from overt hides. Indeed, some military SOF units tasked with high-risk close protection will also employ snipers to provide overwatch and countersniper support, but normally from concealed or covert hide sites. The role of the police sniper in close protection is to provide a protective outer cordon, watching for and capable of interdicting any identified threats.

The Secret Service employs very well trained snipers as part of the protection package they provide to senior government figures including the American president. They will typically deploy from a range of both overt and covert hides, including on-board orbiting helicopters. Along with protecting against sniper attacks, they will support the Secret Service's Counter Assault Team that will conduct the assault against any suspects who threaten one of the Secret Service's "principals" (the term used to describe the person being protected). Intriguingly, when the U.S. president travels overseas to countries that have a heightened security risk, like Iraq or Afghanistan, the Secret Service is supported by an assault troop from either SEAL Team 6 or Delta Force with its own sniper support.

During close-protection operations, the sniper must be vigilant for both gun and bomb attacks against the principal. This means watching for suicide car bombers approaching barriers along with searching for armed individuals in the crowd. They must also perform countersniper overwatch. Indeed, at large events or for exceptionally important people, a number of sniper teams will be assigned specific primary roles.

SNIPER-INITIATED INTERVENTIONS

The types of operations conducted by police and counterterrorist snipers are perhaps the closest to traditional military sniping. In these cases, the sniper typically shoots an individual to stop further loss of life. An example is the infamous 1984 "McDonald's Massacre" in San Diego, which was ended by police sniper intervention. The suspect, a deranged gunman named James Huberty, opened fire on diners with a range of firearms. He murdered some 21 people before being himself killed by a law enforcement sniper over an hour after the incident began.

Two snipers were deployed as part of the cordon as the first responding SWAT team deployed. As the casualties mounted inside the fast-food restaurant, they were both given permission to engage the offender, known today as an active shooter, should he show himself. One of the snipers spotted Huberty and immediately fired a single round that struck him in the upper chest and killed him instantly.

Another example occurred in June 2015, when SWAT officers in Dallas were forced to kill a suspect who had earlier attacked a police station with pipe bombs and rifle fire. He escaped in an armored van that he claimed was rigged with explosives. After negotiations and with the man still threatening to detonate explosives, the vehicle was disabled by .50 sniper fire before the suspect himself was engaged and killed with four .50 rounds.

Police shootings involving snipers are relatively rare within the United States, with suspects more often shot by an entry or a cordon team. They are considerably rarer in other Western countries like the United Kingdom or mainland Europe, where the number of police-involved shootings are a tiny fraction in comparison. Indeed, police sniper shootings are almost unknown in the United Kingdom, despite their prevalence in television dramas. The Metropolitan Police Force's SO19 (now SCO19) Firearms Unit did not have its first sniper shoot until 1988, and they have remained rare since. They are also inevitably at close range.

As an example, one such shooting in October 2000 by a sniper from SO19 was from well under 10 meters (11 yards). The SO19 sniper, using a 7.62x51mm Heckler and Koch G3K, engaged a mentally unstable individual who had begun stabbing a hostage from the window of a neighboring apartment as the entry team smashed their way in. Another operation in 2007 saw a pair of SO19 snipers engage two armed robbers who were threatening the crew of a security van delivering cash to a bank.

These snipers were located only 50 meters (55 yards) from the scene of the incident. The first, armed with a 5.56x45m Heckler and Koch G36K, fearing a security guard was about to be shot, fired one round into a suspect, striking him in the chest and causing him to drop his pistol. His accomplice immediately picked up the pistol and raised it at an SO19 arrest team who were charging toward him. The second sniper fired a single round from his G3K that dropped the suspect to the ground. The sniper fired a second round into him as he struggled with the pistol, killing him.

An early HRT operation involved the kidnapping of a four-year-old boy saw a sniper deployed to end the confrontation and save the hostages. The unit was called in to support a local FBI SWAT team in Sperryville, Virginia in 1988. The suspect was surrounded and, after negotiations that led him to believe the FBI had acquiesced to his demands for a helicopter, he left the farm where he'd been held up. He had the young boy tied to his back and his former girlfriend in front of him as a human shield, a knife at her throat and a pistol in his other hand.

As he neared the helicopter, he appeared about to kill her, pointing the pistol at her head and saying goodbye. A flashbang grenade was lobbed near him to distract him, then a single shot cracked out. A 168-grain 7.62x51mm hollowpoint struck the hostage taker in the side of the head, fired by an HRT sniper located some 64 meters (70 yards) away. The man was killed instantly. Both hostages escaped unharmed.

In 1992, the Bureau's Hostage Rescue Team was deployed to Ruby Ridge, Idaho after a U.S. Marshal Service's Special Operations Group operation against a white supremacist named Randy Weaver, who was dealing in illegally shortened shotguns, went badly wrong. In the initial confused firefight, one of Weaver's sons was shot in the back and killed by the marshals, as was one of the family's dogs. One of the marshals was killed by return fire from Weaver's cabin. The HRT took over tactical responsibility of the incident.

After being issued what can only be described as remarkably loose rules of engagement (that essentially allowed them to fire on anyone carrying a weapon), one of the HRT snipers engaged Weaver as he stepped from the cabin, striking him in the chest with a single round from his .308 Winchester Remington 700. As the wounded Weaver, his daughter and a compatriot raced back inside the cabin, the sniper fired a second round that missed his intended target, punched through the cabin and struck Weaver's wife in the head, killing her instantly as she held the couple's ten-month-old daughter in her arms. The after effects of this shooting clouded the HRT for a number of years and it wasn't really until post 9/11 that the unit's reputation began to improve.

COUNTERSNIPER

Very rarely will a police or CT sniper become involved in a true countersniper operation, although snipers were deployed to attempt to deter the so-called "Beltway Snipers" who terrorized Washington D.C. in 2002. As we've noted earlier, the "Beltway Snipers" actually used a technique seen in both Northern Ireland with the IRA and later by Iraqi insurgent snipers.

In terms of fieldcraft, there are often scenarios in which a police sniper prefers to remain visible as either a deterrent, as in overt hide sites on buildings overlooking the appearance of a VIP, or as psychological leverage for a hostage negotiator. In the latter case, the sniper may well establish his position within sight of the suspect's location to add to the imperative to surrender. This sometimes works with terrorists too. In 1975, an IRA Active Service Unit was cornered in an apartment in Balcombe Street, Marylebone, London. After seeing police snipers from the Met Police's then D11 firearms unit deployed on television, and with the announcement that the SAS had arrived, the terrorists gave themselves up.

There are also situations where concealment is an advantage for police snipers, and they will deploy in Ghillie suits and use elements of military sniper fieldcraft. Consider the hunt for two escaped convicts in New York State in 2015. Snipers from tactical teams from the State Police, the FBI and the Border Patrol were deployed in both static and mobile hides. When police snipers are deployed to gather intelligence on a suspect, they will often use similar techniques to the stalk to infiltrate into the area. Indeed, snipers from agencies such as the DEA (Drug Enforcement Administration) and of course the FBI's HRT are particularly skilled in this area.

AIRCRAFT HIJACKING

Although high in the public imagination thanks to console video games, the storming of a hijacked commercial airliner is the rarest of counterterrorist operations. Although all units train extensively on this type of target, and many, including the British SAS and Australian SASR, have their own airliners that can be configured to match the internal layout of most commercial aircraft, few have ever been called upon to conduct the real-life mission.

Snipers play an essential role in counterhijack operations, both in terms of intelligence gathering and by support of direct intervention against hostage takers. When a hijacked aircraft lands at a major airport, it will be directed to park in a certain area that has been previously identified as able to support a counterterrorism mission, should the decision be made to assault the aircraft to free the hostages. In the United Kingdom, for example, hijacked flights are directed to land at Stansted Airport in Essex, northeast of London. Once civilian and military air traffic control are made aware of a hijacking, a set protocol is followed, which alerts all necessary parties. Key among these is the Essex Police firearms unit.

The initial armed response will be a number of Specialist Firearms Officer (SFO) teams that will immediately throw down an armed cordon around the aircraft, supported by two-man sniper-observer teams. At this early stage, the snipers are the only eyes that authorities have on the aircraft, and their intelligence is quickly fed back to the SFO command post and disseminated to the Home Office, the security services and the military. The Essex Police teams would likely be supported by additional officers from the Met's SCO19, including additional sniper teams. In the 2000 hijacking of an Afghan airliner that landed at Stansted, SCO19 snipers rotated through with military support provided from the SAS.

In December 1994, Air France Flight 8969, which was preparing to fly from Algiers to Paris with 227 passengers on board, was hijacked by four terrorists from the Armed Islamic Group or GIA. While the Algerian government negotiated with the terrorists, GIGN forward-deployed to Spain, within striking distance of Algiers. The terrorists brutally executed one of their hostages when negotiations stalled. The French national counterterrorism unit, GIGN, were authorized to conduct an immediate hostage rescue, one in which their snipers would play a key role.

Before GIGN could deploy to Algiers, the Algerian government allowed the aircraft to take off, its destination Marseilles, France. It can only be surmised that the GIA terrorists were not the most strategic of thinkers, choosing a location that would serve as an ideal staging area for a GIGN assault. Once they had landed, GIGN quickly surrounded the hijacked aircraft, with sniper teams feeding intelligence back to the assault teams.

One of the terrorists opened a side cockpit window and fired his AK at the control tower. This act precipitated the GIGN assault. Using mobile stairways known as air-stairs, the GIGN assaulters breached into three entry points on the aircraft, two at the rear and an eight-man team at the forward door just behind the cockpit. The assaulters carried a mix of 9x19mm Heckler and Koch MP5 submachine guns and their trademark .357 Magnum Manhurin MR73 revolvers.

The terrorists had taken refuge in the cockpit cabin and opened fire on the assault team as they entered the aircraft, following a short struggle to get the cabin door open. A flashbang missed its mark and detonated on the tarmac rather than through the open cockpit side window, although a second flashbang detonated inside the cockpit, temporarily stunning the terrorists. A vicious close-range firefight ensued with several GIGN operators shot and at least one terrorist killed.

The copilot blocked the line of sight of the GIGN snipers, rendering them impotent to intervene to assist the forward assault team. The assaulters at the rear of the aircraft managed to activate the emergency slides and began to evacuate the hostages. The majority of the aircraft had been secured within 90 seconds. While the firefight continued at the cockpit, the copilot who had been blocking the snipers' line of sight managed to escape from the open cockpit window and dropped unceremoniously onto the tarmac below. The snipers began engaging the terrorists with their 7.62x51mm FR F1 sniper rifles, killing two of the terrorists.

As the forward assault element regrouped and pulled their wounded out, the last surviving terrorist opened fire from the cockpit door with his AK. A round struck the weapon of one of the assaulters, sending the operator toppling back down the stairs. Soon after, the last terrorist was shot and killed by a GIGN sniper. In all, ten GIGN operators were wounded, as were a dozen hostages. Incredibly, not a single hostage was killed. All four terrorists died in the operation, three by the intervention of the GIGN snipers.

POLICE AERIAL INTERDICTION

Police and CT units have long practiced aerial sniping for use in counterterrorist missions. The advantages of having a sniper team deployed in a helicopter match those for military SOF: the sniper team can be rapidly moved to where it's needed, it requires no extra time to access a position and, perhaps most importantly, it can pursue, and if necessary, engage, individuals or vehicles. A recent example of police aerial sniping was the deployment of police firearms officers along with RAF Regiment snipers above London during the Olympics.

In some American states, aerial vehicle interdiction (AVI) by police is an approved tactic. These AVIs are not used against terrorists or insurgents, however, but fleeing criminals, some of whom have been accused of relatively minor crimes. Rather than the more militaristic term AVI, these techniques are known as Aerial Use of Force or AUF. In perhaps one of the most notorious of these cases, a sniper on board a Texas Department of Public Safety (DPS) helicopter repeatedly engaged a fleeing pickup truck that contained a suspected drug trafficker.

In the released video of the incident, one policeman even states the sniper is "cleared hot" to try to shoot out the speeding vehicle's tires with a suppressed 7.62x51mm LaRue OBR rifle fitted with an Aimpoint optic and infra-red laser. This is a similar set-up to the tactics used by Delta snipers when they engaged suicide bombers and terrorists in Iraq. In a curious twist, the instructors at DPS who taught the marksman in the helicopter had been trained by Craft International, co-founded by Chris Kyle.

Even after the two rear tires of the pick-up truck were shot and deflated, the pursuit continued and the sniper engaged the vehicle again, blowing out the front left tire. Eventually the truck slowed and stopped. Four illegal immigrants ran from the truck bed. Two others were found in the bed of the truck were announced dead at the scene, apparently struck by the sniper's rounds. The driver hadn't been transporting drugs, but people, under the tarp in the truck's bed. In the wake of the incident, the DPS altered their procedure so that "a firearms discharge from an aircraft is authorized only when an officer reasonably believes that the suspect has used or is about to use deadly force by use of a deadly weapon against the air crew, ground officers or innocent third parties."

Aerial or helisniping is also used by the United States Coast Guard to disable drug smuggling boats, including the famous HITRON or Helicopter Interdiction Tactical Squadron. HITRON fly in Dolphin helicopters, targeting so-called "go-fast" boats, high-speed cigarette boats favored by drug smugglers. They will attempt to stop the suspect boat through the use of audio and visual warnings. If the boat fails to stop, a gunner on the helicopter will fire warning bursts across the bow with the doorgunner mounted 7.62x51mm M240B medium machine gun.

If even this fails to have an impact on the suspect craft, a HITRON sniper from the Precision Marksman Observer Team (PMOT) will engage the boat's engines with the .50 Barrett M107 antimateriel rifle equipped with an EOTech optic, aiming to disable the craft. HITRON had originally been equipped with the bolt-action .50 Robar RC-50, but went to the Barrett to enable faster follow-up shots. They have also recently adopted the 7.62x51mm semiauto Mk14 Mod 0 Enhanced Battle Rifle with the excellent Schmidt & Bender Short Dot scope and the 7.62x51mm M110 variant of the SR-25.

MARITIME HOSTAGE RESCUE

Although not strictly a counterterrorist or law enforcement scenario, we must also discuss the famous rescue of Captain Richard Phillips of the container ship *Maersk Alabama* off Somali waters in April 2009. Four Somali pirates boarded the *Maersk Alabama* and, after encountering resistance from the crew, took the captain hostage in a life boat for four terrifying days. Piracy is big business: in the year prior to the *Maersk Alabama* incident, over 40 container ships had been hijacked and ransomed. The pirates made a mistake targeting the *Maersk Alabama*, however, as it was American flagged.

The USS *Bainbridge*, an Arleigh-Burke class guided-missile destroyer, was the first ship to respond and launched a Scan Eagle unmanned aerial vehicle to provide live video of the hostage taking. The *Bainbridge* requested U.S. Navy SEAL support to end the crisis and an element from Red Squadron of the fabled SEAL Team 6 was dispatched by C-17 cargo plane from the United States. A SEAL Team based in Djibouti flew to the *Bainbridge* to provide a rescue capability until Team 6 arrived.

Apparently the Team 6 operators parachuted into the sea near the *Bainbridge* with their own small boats loaded with equipment. They immediately deployed onto the deck of the destroyer and established sniping positions. The *Bainbridge* carried a Somali interpreter who was instrumental in negotiating with the pirates. With poor weather setting in, the Americans convinced the pirate leader to allow the life boat to be brought under tow behind the destroyer. Incredibly, the pirates agreed. Once under tow, the crew of the *Bainbridge* slowly winched the life boat in closer, unbeknownst to the hostage takers.

After four days held hostage, during an altercation with the ever more unbalanced pirates, one of the gunmen's AK47s was fired into the ocean, apparently as a warning. At that point, the SEALs initiated their deliberate action plan and waited for a clear shot on all three gunmen. When that moment came, three SEAL snipers fired near simultaneously, dropping all three hostage takers with a single headshot each, using their night vision goggles and the LA-5 infrared lasers mounted on their rifles. The reported range was only some 25 meters (27 yards). Other SEAL assaulters boarded the life boat and reengaged the pirates to ensure they no longer presented a threat before safely recovering Captain Phillips.

A variety of different weapons have been attributed to the SEAL Black Team sniper element, from .50 Barretts to the more likely 7.62x51mm SR-25s or Mk11s (and indeed most likely the snipers brought with them a number of platforms to accommodate different shooting conditions). A former SEAL Team 6 operator has, however, confirmed that the lethal shots were in fact made with the 5.56x45mm Heckler and Koch 416 with 14.5 inch (368mm) barrel and equipped with a variable power Nightforce optic and AAC sound suppressor (remembering that their targets were only a scant 25 meters away, the 5.56x45mm was more than sufficient to take the shot.)

POLICE AND COUNTERTERRORISM SNIPER WEAPONS

There is less commonality with police and CT units in terms of their weapons and equipment than with their military counterparts. Although significantly influenced by the weapons used by military

SOF, police and CT units have some requirements that are not matched by military issue weapons.

London Met Police snipers, or rifle officers as they are still known, have for example used a variety of sniper platforms since their inception as D11 and later as PT17 The Force Firearms Unit began life as D11, comprising rifle equipped marksmen who, along with the Diplomatic and Royalty Protection Groups, were among the only regularly armed members of the Metropolitan Police. By 1987 when the unit was renamed PT17, they had assault and sniper sections and were trained to deal with both armed criminals and terrorist gunmen. They were renamed SO19 when they began manning armed response vehicles to cover London in 1991 and have since been renamed CO19 and finally SCO19.

Originally they were equipped with the semiautomatic 5.56x45mm Ruger Mini-14 in this role, but by the 1980s these were replaced by the semiautomatic 5.56x45mm Heckler and Koch 93, a civilian version of the selective-fire HK33. In 1999, the HK93 was supplemented and largely replaced by firstly a modified version of the 7.62x51mm Heckler and Koch G3K with integral bipod and PSG-1 stock, introduced among fears of terrorists wearing body armor, and the 5.56x45mm Heckler and Koch G36K, both with semiautomatic-only trigger groups (traditionally, the majority of police small arms in the UK have had their selective fire settings disabled).

Recently the 5.56x45mm SIG SG516 with a semiautomatic-only trigger group has been adopted by the counterterrorist wing of SCO19. This carbine has largely replaced the MP5 and G36 series as their primary assault weapon, but is also available for short-range sniper coverage. The 5.56x45mm SIG MCX paired with the Aimpoint Micro T-2 red dot sights has also been adopted following rumored UKSF adoption.

France's GIGN use an array of top of the line sniping platforms, including the .338 Lapua Blaser R93 Tactical 2, the .338 Lapua Accuracy International AWM and the 7.62x51mm integrally suppressed Accuracy International AWS and the semiautomatic Heckler and Koch 417, all

POLICE AND COUNTERTERRORISM SNIPING

with Schmidt & Bender MkII variable power optics. When longer-range or penetrative qualities are called for, the unit has both the domestically produced bolt-action .50M2 PGM-50 Hecate II and the semiautomatic Barrett M82A1. For shorter-range situations including covering the advance of assault elements, the semiautomatic 5.56x45mm SIG 550 Sniper with Hensoldt 6×42 BL scope is employed.

Most European units tend toward 5.56x45mm Heckler and Koch G36 series designs for their short- or intermediate-range sniper requirements. Others use a number of SIG designs. For longer-range sniping, the bolt-action platform in 7.62x51mm or .338 Lapua are the leading contenders, often produced by either Accuracy International or the Finnish firm of SAKO. In the United States most units follow the military's lead and use variants of the 5.56x45mm M4 carbine and 7.62x51mm Remington 700 series sniper rifles, often in Accuracy International or McMillan aftermarket stocks.

The Russian tactical units use perhaps the most exotic weapons, rarely seen in the West. Most Russian Vympel and Alfa snipers, for instance, use suppressed 7.62x54mm SVDs, including the folding-stock SVDS variant, the bullpup OTs-03 SVO version and its selective-fire cousin, the OTs-03A SVU-A. The integrally suppressed VSS Vintorez is also often seen in the hands of Russian Federal Police OMON (Otryad Mobilny Osobogo Naznacheniya or Special Purpose Mobility Unit) snipers.

CHAPTER 7
MODERN SNIPER RIFLES

Calibers

Most snipers agree that access to a range of calibers is preferable on the modern battlefield where targets may appear at any number of different ranges. Up to 400 meters (437 yards), the 5.56x45mm is considered sufficient; up to 800 meters (874 yards), the 7.62x51mm is the standard, while out to 1,200 meters (1,312 yards) is the realm of the .300 Winchester Magnum. Beyond 1,200 meters (1,312 yards), both the .338 Lapua Magnum and the .50BMG have their adherents, although the .338 has largely replaced the .50BMG for all but the longest-range shots. We must also consider the different bullet configurations available to the modern sniper. Some types of bullet that are readily available and used by police or counterterrorist teams are not available to the military sniper. These restrictions date back to the Hague Convention of 1899 that banned the use of "bullets which expand or flatten easily in the human body." This has effectively stopped military snipers from using hollowpoint and fragmenting ammunition that, ironically, could be considered more humane, as along with increasing lethality they decrease the risk of overpenetration and injury or death of hostages or civilians.

Because of this restriction, most military sniper rounds are full metal jacketed designs that, as we have seen earlier, tend to create "through and through" wounds.

The problem with such wounds, apart from the danger to noncombatants, is that they will not reliably incapacitate the target. The effect of a bullet upon a human being is partly influenced by psychology. All other things being equal, someone who is already fighting for their life, both physically and psychologically, will be active far longer following a gunshot wound than someone who is not. One former Navy SEAL noted the phenomenon when he described shooting insurgents who were caught unawares versus those he shot while engaged in a firefight. He found the round in question (the 4.6x30mm of the Heckler and Koch MP7A1 sub machine gun) to be more than able against those caught unawares, while a far greater number of rounds were needed to be fired into those insurgents whose fight or flight response had already kicked in.

5.56x45mm

The 5.56x45mm has long been criticized as lacking in both range and lethality. When it yaws correctly, however, even the standard–issue 55 grain M885 can cause devastating injuries. If it doesn't yaw or fragment, it can produce a rather neat hole that, unless it goes through a critical organ, will have little immediate incapacitation effect. Chris Kyle commented on the caliber:"It can take a few shots to put someone down, especially the drugged-up crazies we were dealing with in Iraq, unless you hit him in the head."[1]

The lightweight round is also very susceptible to wind and will often break apart when it strikes an intermediate barrier between itself and the target. Its chief advantages are its low recoil, meaning it's easier to shoot accurately, and its lighter weight compared to the 7.62x51mm, allowing soldiers to carry more ammunition.

During the war on terror, much effort was invested in research to develop bullet designs that would optimize the 5.56x45mm. As a result, a number of designs have been in operational use for nearing a decade that significantly lift the performance of the 5.56x45mm, both in terms of lethality and in increasing the range of platforms chambered for the caliber.

The most common round in use is the 77-grain Mk262 from Black Hills, an open-tip match round designed to optimize the potential of the 5.56x45mm, including in relatively short-barreled rifles. It was originally developed for the Mk12 Special Purpose Rifle to allow accurate hits out to more than 400 meters (437 yards). In Afghanistan, first-round hits have been reported at almost double this. The British are currently working toward their own version of the Mk262 to replace their mediocre L2A2 that was expressly designed not to fragment, thus retarding the lethality of their own round.

The U.S. Marines and some SOF also use the 62-grain Mk318 Special Operations Science and Technology or SOST round that has evidenced similar performance to the Mk262. Some SOF units like Delta use the 70-grain Barnes TSX Optimized Brown Tip in both their intermediate-range sniper platforms and carbines. The Brown Tip was apparently the round that killed bin Laden and thousands of other terrorists and insurgents. A similar round, the 62-grain Mk318 Mod 0, known as the 5.56 Enhanced, is also used by the SEALs.

A number of newer calibers such as the 6.8mm SPC and the .300 Blackout have emerged in recent years as potential contenders. Although these offer some advantages over the 5.56x45mm in terms of range and lethality (the 6.8mm SPC) or performance when suppressed (the .300 Blackout), it remains uncertain whether either will see adoption beyond niche SOF tasks.

7.62x51mm

The 7.62x51mm has been the standard caliber for sniper rifles since the 1950s. Even today, with the increased use of the .300 and .338, it remains the most popular sniper caliber in the world's militaries. The caliber offers acceptable recoil with range capabilities out to 1,000 meters (1,094 yards) in the right hands and generally impressive lethality. Using a much heavier bullet, the 7.62x51mm can suffer from overpenetration, however, which is a consideration for police and counterterrorist units. This can be negated to a degree with recent-generation expanding designs that are meant to expand when they hit and not penetrate through the target. In the U.S. military and numerous Coalition armies, the M118LR is the standard match sniper load, using a 175-grain Boat Tail Hollow Point that often exceeds 1 MOA accuracy with the right platform and shooter. M993 armor-piercing rounds are also available.

Sergeant Dan Mills described the match-grade ammunition British snipers used in their L96A1s:

The rifle takes ball ammunition, exactly the same sort as a normal 7.62mm bullet. The only difference is its green spot ammunition. Green spot is the first batch hot off the presses from a new mould. If you've just made 50,000 rounds, the first 5,000 will be the very best, so they mark them with a little green spot. After a while with metal clunking against metal, small nicks and dents will develop in the bullets. They could minutely affect the round's trajectory in flight. That's why the best stuff is always held over as sniper ammunition.[2]

.300 WM

The .300 Winchester Magnum has long been a favourite of U.S. SOF units. As we will detail later, even the standard-issue 7.62x51mm M24 sniper rifle of the U.S. Army was originally designed to be able to accept the larger round. The .300 offers increased range, velocity and lethality over the 7.62x51mm with little increase in recoil or ammunition weight.

The Mk248 Mod 1 is the standard U.S. sniper load for the .300 Winchester Magnum, a 220-grain boat-tail hollowpoint design developed by Crane Division of the U.S. Navy's Naval Surface Warfare Center to extend the lethality of the caliber. The Navy SEALs had been asking for an improved round for their Mk13s that could push the round beyond its nominal 1,200-meter (1,312-yard) effective range.

The Crane engineers managed to wring every ounce of performance from the cartridge and added an additional 170 meters (185 yards) to its engagement envelope, for a 1,370-meter (1,498-yard) capability. The new round was additionally designed to decrease the effects of crosswinds, and decrease muzzle flash, admittedly less of an issue when deployed with the Mk11 suppressor as the Mk13 customarily is.

.338 Lapua Magnum

Chris Kyle favored the .338 Lapua as he felt it "shoots farther and flatter than the .50 caliber, weighs less, costs less, and will do just as much damage."[3] The development work that ended in the .338 Lapua was originally commissioned for the U.S. Marines, who were looking for a longer-range sniper caliber. Its final design was thanks to a joint endeavor by rifle manufacturers Accuracy International and Sako and the ammunition producer Lapua.

A number of European nations and SOF units saw the advantages of the round and adopted it, but it was not until the British Army phased out their 7.62x51mm L96A1 platforms in favor of the .338 Lapua L115A3 that the round became widely popular. With recent combat experience in Afghanistan, the .338 Lapua offered the perfect solution to long-range shooting in arid conditions.

Another more recent design is the .338 Norma, which has been adopted by a handful of SOF units and has been developed to make the most of the 300-grain boat-tail hollowpoint round, along with reducing the wear and tear on match-grade barrels in comparison to the .338 Lapua. It has also been offered as an option to increase the range of the M240 medium machine gun.

.50BMG

The .50BMG has been somewhat eclipsed by the .338 Lapua in recent years, but for truly extreme long-distance shooting, it is hard to beat.

Although punishing in its recoil, the .50BMG offers an incredibly lethal round out to extremely long ranges. The Raufoss Mk211 Mod 0 is the most common .50BMG round in use today. Based on a Norwegian design, the Mk211 offers impressive accuracy along with increased lethality.

The Mk211 makes the most of the accuracy potential in the M107 that is normally rated as 2.5 to 3 MOA. With the Mk211 it moves closer to a 2 MOA or better platform, at least under 1,000 meters. It is built around a tungsten carbide penetrator, making it capable of penetrating light armored vehicles and most common cover. The round also features an incendiary mix in the nose backed by a small amount of high explosive, making it supremely multipurpose.

Tests have indicated that the round tends to have already penetrated through a human body before the explosive element detonates, but anecdotally there are numerous tales of targets exploding when struck by the round. It also appears that the explosive has a useful effect against other targets who may be standing near the initial target when it detonates, as the round tends to overpenetrate through the original target and then explode, causing fragmentation wounds to those closest the target. The Red Cross have actually argued for the banning of the round's use against human targets.

BOLT ACTION OR SEMIAUTOMATIC

The argument around the relative merits of the bolt action versus semiautomatic date back to World War II, when the first semiautomatic designs began to appear, such as the German 7.92x57mm Gewehr 43. The conventional wisdom is that the bolt-action design causes less movement in the weapon's receiver, as there is only one stage of recoil, and is thus more inherently accurate. A semiautomatic typically has three distinct recoil movements when fired, the round exiting the barrel, the bolt flying backward to eject the spent casing and finally the forward bolt motion as it strips another round from the magazine.

There is also a very valid argument that the bolt action is a far simpler design with fewer things to go wrong. The bolt action is also much less prone to stoppages, as the reasons for stoppages with a semiautomatic weapon simply don't exist with the bolt action. Bolt actions can of course still suffer a stoppage, but it is far more likely to be an ammunition quality issue. Proponents of the semiautomatic say that with modern designs and good maintenance this is far less of an issue than it may once have been.

Many combat veterans argue that the questionable improvement in accuracy, particularly with modern free-floating barreled designs, is less important than the ability to deliver a fast follow-up shot or to engage large numbers of enemy. Although the bolt action may still reign supreme, the gap between the bolt action and the semiautomatic sniper rifle designs has narrowed considerably.

The author asked Sniper Master Nathan Vinson his view.

Accuracy was, and still will remain, our snipers' priority over firepower: however, current and past deployments have seen teams caught out and killed due to lack of firepower or a semiautomatic weapon platform. Initially we advised the teams to deploy [to Afghanistan] with two weapons i.e. a primary and secondary [not including a pistol]: however, this at times proved too much additional weight and equipment to carry. So to minimize [the] load, the HK417 was introduced to sniper teams and platoon marksmen across the Australian Army. The SR98 was however still used on a number of missions, but close support for those teams was close by.

Vinson explains the range of weapons typically available to Australian Army sniper teams, both bolt action and semiautomatic. "The teams will have a choice of four weapon platforms. The first is our SR98 7.62mm bolt action; second is the HK417 7.62mm semiauto; thirdly [the] Blaser .338 [bolt action]; and finally the AW50 .50cal bolt action. Any one of these weapons can be used but [that] will depend on the target, mission and the effect the sniper team is to impact on the target or target area." Similarly, in the U.S. Army and Marines both bolt-action and semiautomatic platforms are available.

In Afghanistan, the preferred weapons carried with my boys were the HK417 and Blaser .338 bolt action. The HK417 because of its firepower and accuracy and the Blaser .338 because of its accuracy, range and it's a damned sight lighter than the AW50! Previous teams were also issued the SR25 7.62mm semi, but by the time we deployed the weapons were on their last legs.

He has also seen the advantage of a semiautomatic but sees a place still for the bolt action, and raises a particularly interesting point about prone shooting positions that are unobtainable because of the design of the semiautomatics; the location of the pistol grip and magazine means that the sniper cannot get as low to the ground, literally perpendicular, as they can with most bolt-action designs.

I first trained on and deployed with the old Parker Hale 7.62 bolt action then later the SR98. But I would say in the recent years I have now converted to the HK417 for rapid engagements of multiple targets and the Blaser just to punch out a little further. Saying that, I still love the feel of the bolt action on the SR98, the feel of its recoil and the ability to be able to remain closer to the ground when taking a shot, which more often than not you cannot achieve with the semiautomatics.

BOLT ACTION RIFLES

M24

Caliber: 7.62x51mm
Unloaded Weight: 6.04kg
Overall Length: 1,035mm
Range: 800m+
Origin: United States

The U.S. Army's M24 Sniper Weapon System (SWS) is a 7.62x51mm militarized version of the venerable civilian Remington 700. It was first adopted in 1988 after extensive and lengthy testing that saw the Remington go up against a bevy of the best rifles of the time: the U.S. Marines' M40A1, the French FR-F1, the commercial Winchester 70 and the Parker Hale M82 (later adopted by the Australian and Canadian armies, among others).

Apparently members of the U.S. Army Special Forces were involved in the selection of the M24, as they wanted a platform that could outshoot the then current issue 7.62x51mm M21 in terms of distance and accuracy. They also wanted a heavier round than the 7.62x51mm, and they specifically identified the .300 Winchester Magnum as their preferred option.

Unfortunately, the Army could not at the time decide on adoption of a non-NATO caliber without extensive testing that would slow down M24 procurement by an estimated four years. The Green Berets relented on their .300 Winchester Magnum requirement, but asked for a long action in the M24 that would be capable of rechambering for the larger caliber in the future.

The M24 is commonly equipped with the Leupold Mark 4 10 power optic and a Harris folding bipod to improve stability. The M24 has a 24-inch (660mm) free-floating barrel with a 1 in 11.25 inch twist. The later M24A2 features a threaded barrel to accept the issue suppressor. Showing its civilian heritage, the M24 features an internal five-round magazine. The A2 version switched to a detachable ten-round magazine. The stock on all three versions can be adjusted to suit the individual sniper.

The M24 has an effective range of up to 800 meters (874 yards). However, as several earlier accounts in this book have shown, it can certainly still be effective beyond this range.

It is also an exceedingly accurate weapon, easily capable of 1 MOA accuracy with the match M118 or sub 1 MOA with the M118LR round.

The M24 has most likely been responsible for the deaths of more enemy during the wars in Afghanistan and Iraq than any other, a testament to its solid, dependable design. Its successor, the M2010 that we will detail in a moment, builds on that heritage, essentially updating and upgrading a classic sniper rifle to match the operational requirements of the 21st century.

M2010

Caliber: .300 Winchester Magnum
Unloaded Weight: 8.07kg
Overall Length: 1,122mm
Range: 1,200m+
Origin: United States

The U.S. Army officially describes the new M2010 as follows: "enables sniper teams to engage enemy personnel with an M24 Sniper Weapon System converted to fire .300 Winchester Magnum (Win Mag) ammunition. It offers greater precision and a 50 percent increase in effective range beyond current Army 7.62mm sniper rifles."

Combat experience in Afghanistan had shown the U.S. Army that a heavier standard sniper caliber than the 7.62x51mm was preferable. In what snipers referred to as the "ridgeline to ridgeline fight," the 7.62x51mm of the M24 and the M110 weren't providing the necessary reach to engage enemy targets at extended ranges, particularly at altitude with high- and low-angle shooting. Further, they wanted a system that was lighter and handier than the .50BMG M107, which was a very heavy weapon to carry in the mountains, but that could still make hits beyond the 1,000-meter (1,094-yard) mark.

Building on the positive operational reputation of the SOCOM Mk13 platform and the .300 Winchester Magnum caliber, an Army requirement for an upgrade to the M24A2 was announced in 2009, with Remington winning the contract the following year. Fielding began to units deploying to Afghanistan in mid-2011. The new platform was christened the M2010 Enhanced Sniper Rifle or ESR.

The issue M24A2 was selected as the basis for the new rifle, as an upgrade was considered the faster way to fielding the new platform. It also reduced training time, as the mechanics of the weapon would function in a very similar way. The long-action receiver of the M24A2 had always been designed with larger calibers in mind, and the modifications required to rechamber the M24A2 to the .300 Winchester Magnum were comparatively straightforward, although the only element of the venerable M24 to survive the upgrade was the receiver. A new 24-inch (609mm) free-floating barrel with a 1 in 10 twist was fitted, meaning the round was very stable by the time it exited the barrel.

The decision to purchase the long-action version back in the 1980s had been an uncharacteristically smart call by the Army, and the dividends paid off over 20 years later with the M2010 (while the Marines, as we shall shortly see, were stuck with the short-action M40). Along with the caliber, the biggest change was the new chassis, with a user-customizable side-folding stock. The stock can be adjusted for length and height to ensure the best cheek weld for the sniper. One Army sniper said: "the weapon system is probably the most user-adjustable system in the Army. You can practically adjust everything on the gun to fit you as a person and how you like to set it up."

The M2010 has been fitted with a full-length Picatinny rail that runs along the top of the weapon, allowing for the easy attachment of night vision devices. The M2010 is also issued with an Advanced Armaments Corporation Titan QD Fast Attach suppressor, and mounts the Leupold

Mark 4 6.5 to 20 power variable optic that includes a Bullet Drop Compensator calibrated for the issue Mk248 round. The optic features the Horus H-58 reticle that simplifies and vastly speeds up elevation and windage corrections, allowing for much faster engagement times (see Appendix 1 for a former special operations sniper's view of the Horus).

The M2010 has been very well received by U.S. Army snipers in Afghanistan, who appreciate the extended range and lethality of the weapon and the .300 Winchester Magnum bullet. The adoption required some special planning when units transitioned to the M2010. At Bagram Air Base, there were no ranges that could handle targets beyond 1,200 meters (1,312 yards) so makeshift ranges were built "outside the wire," requiring infantry patrols to first clear and secure the area before the snipers could begin their transition training.

M40

Caliber: 7.62x51mm
Unloaded Weight: 6.04kg
Overall Length: 1,035mm
Range: 800m+
Origin: United States

The U.S. Marine Corps adopted the Remington M40-based 7.62x51mm in 1966 and the weapon served as the principal sniper rifle for the Marines during the Vietnam War (although the semiautomatic M14 and M21 were available in limited numbers). The M40 has been constantly upgraded over the 50 years of its service, so much so that the latest version, the M40A5, hardly resembles the original.

In 1977 the M40 became the M40A1, with its traditional wooden stock replaced with a fiberglass stock manufactured by McMillan, who would later become a major player in the sniper world. In 2004, the M40A3 was issued, which added a heavier match-grade stainless steel barrel and another McMillan stock. For most of its extended service life, the M40 platform has been matched with the fixed 10-power magnification Unertl. From the M40A3 variant, the 3 to 12 power Schmidt & Bender M8541 was issued.

M40A5 is current USMC issue, although the Marine Scout Sniper community has been arguing for a heavier caliber option after Afghan experience despite the fact that Marine snipers have been making the most out of their M40A5s, registering kills out to 1,000 meters. Instead of the Mk21 or M2010, the Marines are looking to issue the restocked M40A7 with a skeletonized stock to reduce weight but still chambered for the 7.62x51mm round. Unfortunately the M40 uses the Remington 700 short-action rather than the long-action variant, like the Army M24, which allows for reasonably simple rechambering to a larger caliber like the .300 Winchester Magnum.

A Marine Corps spokesperson said in 2015:

> We have made engineering changes to the M40A series in order to modernize the weapon. Our efforts focused on maintaining its relevance. Within the next year, we plan to field the M40A6 which will incorporate a modular stock to improve portability and shooter ergonomics, an improved barrel and an upgrade of the ballistic calculator to reduce weight and improve accuracy.
>
> The Marine Corps does have a requirement for a precision-engagement capability that exceeds the M40. We continue to pursue a common solution with the Army and USSOCOM and will explore all available options.

Another Marine commented on the ongoing controversy: "We make the best snipers in the world. We are employed by the best officers in the military. And we are the most feared hunters in any terrain, but the next time we see combat, the Marines Corps is going to learn the hard way what happens when you bring a knife to a gunfight."

Surprisingly, former Ranger sniper Sergeant Nicholas Irving chose the M40A5 in 7.62x51mm as his favorite sniper rifle, along with his much-loved suppressed SR-25. Irving explained that working with Marine Scout Snipers in Helmand had given him a good insight into the veteran sniper rifle's capabilities. Indeed with the M118LR match ammunition, .5 MOA shooting is not uncommon with the M40A5.

Mk13

Caliber: .300 Winchester Magnum
Unloaded Weight: 5.17kg
Overall Length: 1,206.5mm
Range: 1,200m+
Origin: United States

In the late 1980s, U.S. SOCOM requested a heavier variant of the M24 firing the .300 Winchester Magnum round, apparently as part of a SEAL requirement. The result was the Crane-developed Mk13 Mod 0. It has since progressed through a number of variants, with upgrades to the optics and chassis of the weapon. It uses the Mk11 suppressor for commonality across weapon types in both .300 Winchester Magnum and 7.62x51mm.

The Mk13 has served with distinction with the SEALs and the Rangers in particular, where the platform was extensively deployed by Ranger sniper teams in Afghanistan, with two-man teams using the SR-25/M110 and the Mk13 or "300 WinMag," as it was known. The Mk13 was even considered as an interim purchase for the U.S. Army to narrow the capability gap between the 7.62x51mm M24 and the .50BMG M107, and was extensively tested during 2009. Ultimately an upgraded .300 Winchester Magnum version of the M24, the M2010, was selected instead of purchasing an entirely new platform.

Chris Kyle largely favored the Mk13, particularly the Mk13 Mod 5, which featured an AICS (Accuracy International Chassis System) folding stock. Indeed, most of his kills were with the Mk13. U.S. Marine MARSOC snipers in Afghanistan have reportedly registered hits out well beyond 1,500 meters (1,640 yards) and apparently as far as 2,000 meters (2,187 yards) using the latest iteration, the Mk13 Mod 7, along with the latest Mk248 match-grade ammunition.

Mk21 PSR

Caliber: 7.62x51mm, .300 Win Mag, .338 Lapua Magnum
Unloaded Weight: 7.7kg
Overall Length: 1,200mm
Range: Variable
Origin: United States

The PSR or Precision Sniper Rifle is almost the U.S. SOCOM version of the Enhanced Sniper Rifle or ESR program that resulted in the adoption of the M2010 for the U.S. Army. U.S. SOCOM snipers wanted a multicaliber-capable, sub 1 MOA rifle that could be quickly and easily switched out between 7.62x51mm, .300 Winchester Magnum and .338 Lapua Magnum. As we have noted earlier, this was a direct effect of operational experiences in Afghanistan. The PSR could in effect be rechambered in the field for a more suitable caliber, but this would more likely occur at base when the mission was briefed.

A number of internationally renowned firms built prototypes for the competitive trial. Accuracy International, Remington, Barrett and Desert Tec (unusually with a bullpup design that placed the magazine behind the receiver) built state-of-the-art prototypes, but the Remington MSR or Modular Sniper Rifle design won the contract. The PSR was designed from the ground up for use with a detachable suppressor, and the final design incorporates a suppressor that can be used across all three calibers.

The weapon has been deployed to Afghanistan, and now most likely Iraq, but the platform is too new for any combat reports to have filtered out at the time of writing. However, the author has seen a 2016 video from Northern Iraq that shows what appears to be a Mk21 in the hands of an unidentified U.S. Army special operator, embedded with Kurdish Peshmerga, taking shots at static Islamic State forward positions with the new rifle. Intriguingly, however, SOCOM have issued a request for the development of an armor-piercing .338 Lapua round for deployment with the Mk21, with the specific requirement of being able to defeat Level IV body-armor plates at ranges out to at least 400 meters (437 yards).

Accuracy International Arctic Warfare (L96A1, L118A1, SR-98, G22)

Caliber: 7.62x51mm, .300 Win Mag (G22)
Unloaded Weight: 6.5kg
Overall Length: 1,180mm
Range: 800m+
Origins: United Kingdom

The British Army had used the 7.62x51mm L42A1, a rebarreled Enfield design, since the early 1960s. Even before the Falklands War in 1982, where British snipers were outgunned by Argentine snipers equipped with more advanced night vision scopes, along with better rifles in the SGG-69 and Weatherby .300 Winchester Magnums, it was abundantly clear that a more modern replacement was required.

The story of that replacement owes much of its later success at the British Army trials to both its chief designer, the late Olympic target shooter Malcolm Cooper, and to the UK Special Forces. Cooper established Accuracy International (AI) in 1978 with engineers and competition shooters Dave Walls and Dave Caig. Their original designs were aimed squarely at the civilian shooter, but a new design, known as the Precision Marksman (PM), was developed from the ground up as a military sniper rifle.

The firm recognized that until that time most sniper rifles, like the U.S. Army's M24 and the U.S Marines' M40, were militarized versions of commercial hunting or target rifles. Instead, Accuracy International actively sought input from British Army snipers in the development of the PM. By 1984, prototype examples were being tested by the Special Boat Service (SBS) who purchased the first eight from the production line a year later. The SAS also liked what it saw and 22 SAS procured 32 of the rifles. Accuracy International, and the PM, were off to a flying start.

Learning of a British Army requirement for a replacement for the L42A1, Accuracy International developed another prototype based on the PM, but with refinements suggested by their SBS customers and a new lightweight stock. The prototype was consistently registering 85 percent first-round hits out to 900 meters, and in 1985 the AI design beat a number of other superlative sniper rifles such as the Heckler and Koch PSG-1, the SIG Sauer SSG 2000, the Parker Hale M85 and the Remington 700 to win the contract for the new British Army sniper rifle, the L96A1.

The L96A1 was designed, according to the British Army, to be able to hit man-sized point targets with its first round out to 600 meters (656 yards) and to provide harassing fire out to 1,100 meters (1,203 yards). It was issued with the Schmidt & Bender 3 to 12 variable magnification PM II scope and a detachable sound suppressor. In terms of accuracy, the platform is outstanding, consistently achieving sub 1 MOA with military match-grade ammunition.

Sergeant Dan Mills commented on the L96 he used in Iraq in his classic book Sniper One: "The L96 is accurate enough to kill at 900 meters, and harass up to 1,100 meters. The restrictions are not human nor the fault of the rifle. After that sort of distance, the 7.62mm round just hasn't the explosive charge powerful enough to allow it to fly any further. It then starts to drop out of the sky."[4]

An upgraded version of the L96A1, the Arctic Warfare, was sold to the Swedish Army and eventually replaced the L96A1 as the L118A1. It was adopted by the Australian Army with the optional side-folding stock as the SR-98 and was purchased in some numbers in both suppressed and non-suppressed configurations by Delta Force. Indeed, the AWS or Arctic Warfare Suppressed model was also very successful with a number of prominent European counterterrorist units, including the Italian GIS (Gruppo di Intervento Speciale), thanks to its excellent sound reduction particularly with subsonic ammunition. With the suppressor comes an unavoidable reduction in range, though, and the AWS is rated as effective out to only 300 meters (330 yards). Intriguingly, however, the sniper can replace the suppressed barrel with a standard AW version in minutes.

Another version, the AWC or Arctic Warfare Covert, also proved popular with various SOF including the German Army special operations unit, the KSK (Kommando Spezialkräfte), who adopted it as the G25. The AWC is a real James Bond-style breakdown sniper rifle that is supplied in its component parts in a suitcase. The AWC featured a compact 12-inch (305mm) barrel with integral suppressor and a side-folding stock to further reduce its overall length. A similar rifle, although semiautomatic in nature, was recently adopted by JSOC for their Clandestine Break-Down Rifle or Clandestine Sniper Rifle, although it is unknown exactly which rifle was adopted as the trials were shrouded in understandable secrecy.

The Germans also adopted a version of the AW in .300 Winchester Magnum known as the G22, with an effective range of 1,100 meters (1,203 yards). In the late 1990s, a variant of the Arctic Warfare Model known as the SR-98 replaced the Parker Hale M82 in Australian service. With Schmidt & Bender optic and issue suppressor, the SR-98 has served with distinction in East Timor, Afghanistan and Iraq. The AW is also the standard sniper rifle used by the Spanish military in both Afghanistan and Iraq (intriguingly, they also used the bolt-action variant of the .50BMG Barrett, the M95, along with a bipod- and optic-equipped 5.56x45mm Heckler and Koch G36 to arm their spotters).

As this book went to press, the German Army announced the adoption of a new bolt-action sniper platform to supplement the G22 in the hands of German SOF such as the KSK and KSM known as the G29 in .338 Lapua Magnum. The G29 is a modified version of the civilian Haenel RS9 precision hunting rifle. Intriguingly, the G29 is issued with an Aimpoint Micro red dot sight that sits above the main Steiner optic to facilitate close-range shooting and, like the G22, features a side folding stock.

Accuracy International Arctic Warfare Magnum (AWM, L115A3)

Caliber: .338 Lapua Magnum
Unloaded Weight: 6.9kg
Overall Length: 1,230mm
Range: 1,500m+
Origin: United States

The British Army and Royal Marines adopted the L115A3 in 2007, rushing the first examples into the field in Afghanistan as their snipers transitioned from their 7.62x51mm L96A1s and L118A1s to the heavier .338 Lapua platform. A version known as the L115A1 had been in service with UK Special Forces, who had been using the rifle for some time with much success. It was UK Special Forces' positive experiences with the platform in Afghanistan that influenced the 2007 adoption. A number of American SOF, including Delta Force, had already combat trialed and adopted the Accuracy International .338 Lapua based on UK Special Forces' use.

Veteran sniper Sergeant Craig Harrison recounts in his book the first time he was issued the L115A3 in Helmand:

Simplistically, the .338 rifle is a bigger, heavier version of the L96, shooting the larger .338 bullet. The L96 has an effective range of 900 meters and can harass out to 1,100 meters, whereas this bad boy can be comfortably used at 1,500 meters plus. It has a more powerful scope and a much more powerful round. It was like Christmas for Eddy and me. We just couldn't stop playing with it.[5]

The L115A3 adopted by the British Army and Royal Marines uses a Schmidt & Bender variable 5 to 25 power PM II scope. It includes an adjustable cheek piece and butt stock that can be shortened or lengthened by the use of spacers. The L115A3 has a fluted barrel, meaning that it has a number of grooves on the outside of the barrel that assist with cooling, a system that is commonly seen on the M107 and many heavier caliber sniper and antimateriel rifles.

In terms of range and accuracy, the L115A3 is intended for use out to 1,500 meters (1,640 yards), but as we have seen in earlier chapters, British snipers have been consistently engaging insurgents at well beyond this range. Rated to achieve first-round hits out to 1,100 meters (1,203 yards), the L115A3 is a sub 1 MOA platform, indeed regularly achieving .5 MOA during service in Afghanistan.

The Accuracy International AWM in .338 Lapua have been purchased in small numbers by Russian special operations, and surprisingly have turned up in news photos from Syria in the hands of Syrian SOF. At least one example, fitted with a suppressor and hand-painted in desert tan, has been photographed. Versions have also been in use with certain American SOF units for much of the last decade, including a special 20-inch (508mm) barreled compact version.

A new version of the Accuracy International platform in both 7.62x51mm and .338 Lapua has been developed as the AX, a version of which was a recent contender for the U.S. SOCOM Mk21 Precision Sniper Rifle. The stock of the AX, known as the AICS or Accuracy International Chassis System, is now available for Remington 700 systems, much like the earlier AI stocks used on the Mk13.

C14 MRSWS (Medium Range Sniper Weapon System)

Caliber: .338 Lapua Magnum
Unloaded Weight: 7.1kg
Overall Length: 1,200mm
Range: 1,500m+
Origin: Canada

The C14 is the designation given by the Canadian Army to its replacement for the C3A1 sniper rifle that served since the 1970s, itself a version of the venerable Parker Hale M82. The C3A1 was chambered for the 7.62x51mm and featured a USMC-style Unertl 10 power scope mounted upon a McMillan stock.

Along with many NATO nations, and building upon their own experiences in Kandahar Province, the Canadian Army procured a replacement for the 7.62x51mm design in the form of the .338 Lapua Magnum C14 Timberwolf manufactured by Prairie Gun Works, a small Canadian firm. The rifle has served alongside the C3A1 as the latter was slowly replaced.

The new rifle features a free-floating match barrel and the common Leupold Mk4 variable magnification optic to achieve a reported .7 MOA level of accuracy. Intriguingly, the British SAS have been seen in action with the rifle in Helmand, and although there has been no official confirmation, the manufacturer does list the UK Ministry of Defence as a client. In a curious twist of fate, a number of the same type of Timberwolves (and at least one suppressed .50 PGW LRT-3) originally supplied to Saudi SOF have been captured and displayed by Houthi insurgents in Yemen.

DSR-1

Caliber: 7.62x51mm
Unloaded Weight: 5.9kg
Overall Length: 990mm
Range: 800m+
Origin: Germany

The DSR (Defensive Sniper Rifle), one of the world's most expensive sniper rifles, is employed by a handful of Western European counterterrorist units including GSG9 and the Spanish Army special operations unit, the GEO (in fact it was apparently developed with input from GSG9). Along with being one of the most expensive, it is also one of the most accurate sniping rifles ever produced, with accuracy reported to within .2 to .5 MOA. Unusually for a sniper rifle, it is a bullpup design, meaning the magazine and receiver are situated behind the pistol grip.

The weapon features a monopod in the stock and almost every feature of the rifle is fully adjustable. There is even a space for a spare magazine to be mounted on the receiver. Unusually, the integral bipod is not so much attached to the weapon, but the weapon attached to the bipod. This eliminates any effect of the bipod on the free-floating barrel. Users report the muzzle blast to be dramatic, but the recoil, thanks to the weight of the weapon, is very manageable.

Uniquely, the DSR's suppressed version, the DSR-1 Subsonic, features a suppressor unit that attaches directly to the body of the rifle rather than the barrel, thus eliminating any impact on accuracy from the weight of the suppressor on the free-floating barrel. The DSR-1 .338 Lapua Magnum is also available with conversion kits to rechamber for the use of 7.62x51mm and .300 Winchester Magnum.

Blaser R93 Tactical 2

Caliber: .338 Lapua Magnum
Unloaded Weight: 5.8kg
Overall Length: 1,229mm
Range: 1,500m+
Origin: Germany

Another German design, although this time originally developed from a sporting weapon, the Blaser series is known for its superb accuracy, with .5 MOA or better being the standard, even in combat conditions. They are also well known for their unique straight-pull bolt. This, as the name indicates, is a bolt mechanism that is manipulated straight back without any cant, leading to much faster reloading of the bolt action.

Australian SOF has deployed the weapon, often with a suppressor, with much success to Afghanistan, and it is used by a number of European counterterrorist and police tactical units. A conversion kit is available that rechambers the weapon to 7.62x51mm. It can also be quickly and easily broken into two components and reassembled with apparently no loss of zero.

Sako TRG-42

Caliber: .338 Lapua Magnum
Unloaded Weight: 5.3kg
Overall Length: 1,200mm
Range: 1,500m+
Origin: Finland

The Finnish firm of Sako is recognized as one of the finest manufacturers of precision rifles in the world today. Its TRG-42 chambered for the .338 Lapua Magnum is used by many European counterterrorist and tactical police units, along with SOF from France, Denmark, Sweden and Italy. The Spanish national counterterrorist unit, GEO, still uses the earlier model Sako TRG-41 along with an integrally suppressed variant, the A-II, while the Italian GIS uses the more modern TRG-42.

Typical of Sako rifles, it features a cold hammer forged receiver, free-floating barrel, an adjustable trigger, integral bipod and adjustable stock for best cheekweld and to accommodate differing types of body armor. Unusually for a bolt-action sniper rifle, the TRG-42 features folding iron sights should the optics become damaged.

SEMIAUTOMATIC RIFLES

In this section we will cover both semiautomatic sniper rifles and Designated Marksman Rifles or DMRs. To separate the two classes is difficult as there is significant overlap, with snipers often employing platforms also used as DMRs and vice versa (although the greatest example of this is probably the employment of the British bolt-action L96A1 as a Platoon Marksman's rifle).

SR-25 (Mk11 Mod 0/1 and M110)

Caliber: 7.62x51mm
Unloaded Weight: 4.8kg
Overall Length: 1,118mm
Range: 600m+
Origin: United States

The SR-25 was developed in the early 1990s at the request of Delta Force, who were looking for a 7.62x51mm semiautomatic sniper platform after combat experiences during Desert Storm. At the time, Delta were still using the 7.62x51mm M14 but wanted a more modern AR design. Knight's Armament Corporation worked closely with Delta over a number of years to refine the weapon based on operator feedback. The SR-25 was based on the original AR-10 design but upgraded, with a free-floating barrel and the addition of Picatinny rails among a host of other improvements. SEAL Team 6 and the Rangers also adopted the SR-25 in limited numbers.

A requirement from the SEALs through U.S. SOCOM produced a version of the SR-25 from Crane that was eventually type-classified as the Mk11 Mod 0 in 2000. The Mk11 saw immediate and successful service with the SEALs, Rangers and Green Berets deploying to Afghanistan. The Mk11 was deployed as both a sniping and DMR platform by the Rangers for many years, along with limited service within conventional units. When the U.S. Army was looking for a semiautomatic 7.62x51mm sniper rifle, the Mk11 was an obvious contender.

The 7.62x51mm M110 SASS or Semiautomatic Sniper System, adopted in 2007, is based on the Mk11 and incorporates upgrades made to both current specification SR-25s and Mk11 platforms. Like its predecessors, the M110 is an evolution of the AR-10 platform, feeding from a detachable 20-round magazine. Unlike the M24, the M110 is equipped with a variable-magnification scope, in this case the Leupold Mark 4 3.5 to 10 power with illuminated reticle.

To allow straightforward changes of optics, the Leupold's scope mounts attach directly to the Picatinny rail over the weapon's receiver. Should the rifle's optics become damaged, the scope can be quickly detached and the integral folding backup iron sights can be used. Like the M24, the SASS also features a folding Harris bipod. It comes factory-threaded for the issued suppressor.

The SR-25, Mk11 and M110 all have an effective range of some 1,000 meters, according to the manual, and shoot between 1 MOA and 2 MOA depending on ammunition. Its big advantage for the military sniper (and spotter) is its ability to transition to an assault rifle. One U.S. Army sniper commented that "I don't have to have my shooter carry an extra weapon when we go into buildings to clear rooms. He can actually use (the M110). That's going to lighten our load a lot."

Mk12 Mod 0 and Mod 1

Caliber: 5.56x45mm
Unloaded Weight: 5.5KG
Overall Length: 946MM
Range: 400M+
Origin: United States

What is today known as the Mk12 Mod 0 and Mod 1 Special Purpose Rifle (SPR) began life as a Delta Force initiative of the 1990s to develop a light sniper rifle that could be used for intermediate-range shooting and that could be used for CQB. Particularly after their experiences in Mogadishu, Delta's recce snipers developed the first Recce Rifles in conjunction with Knights Armament Corporation (in much the same way that the first SR-25s had been developed).

According to former Delta operator and firearms trainer, Larry Vickers, these early 5.56x45mm rifles mounted a Japanese 4-power Microdot optic that featured a red dot along with the standard reticle. This was replaced by the Leupold CQT and later the Schmidt & Bender Short

Dot, both produced in association with the Army unit. Early models were fitted with free-floating SR-25 forward handguards to provide increased room for mounting lights and lasers.

While Delta experimented with various incarnations of the so-called Recce Rifle, SEAL Team 6 were exploring similar concepts. A formal requirement for such a weapon was raised at Crane and work began. The original Mk12s were built from M16A1 lowers matched with a 16-inch barrel (the original Mk12s mainly had 18-inch barrels, but these were often swapped out when they arrived at SEAL Team 6) and fixed M16A1-style stock. They were apparently capable of .5 MOA shooting out to 600 yards (548 meters), phenomenal shooting from a carbine platform.

Soon other SEAL Teams were seeing the versatility of a lightweight 5.56x45mm that could reach targets at extended ranges, and the Mk12 began to be issued across the Naval Special Warfare community from 2003. It was also issued across a number of Army units, including the Rangers and Green Berets. The U.S. Marines also deployed the weapon as an SDMR mounting the Leupold 2.5 to 8-power Mk4 scope. With the Mk262 ammunition, Marine SDMs have recorded center of body mass hits at 800 meters (874 yards) and headshots out to 600 meters (656 yards).

Chris Kyle commented that he found the SPR to be very easy to handle, thanks to the weapon essentially having the same controls as the M4 and M16. He also modified his Mk12 by replacing the lower receiver with one from an M4A1, giving him a fully automatic capability, although he never used it in combat. He replaced the standard M16 stock with a collapsible M4 type, although this sometimes caused the weapon to be unreliable when fired on full automatic. The Mk12 has featured prominently, and accurately, in the film Lone Survivor, with two of the SEALs carrying the 5.56x45mm Mk12 Mod 1. The other two members of the reconnaissance team carried the M203A1-equipped M4A1s sporting ACOG optics.

Mk14 Mod 0/1 (M14 EBR, M39 EMR)

Caliber: 7.62x51mm
Unloaded Weight: 5.1kg
Overall Length: 889mm
Range: 600m+
Origin: United States

The Mk14 was another platform that was first designed for the U.S. Navy SEALs. The SEALs had been using a number of versions of the venerable M14, including the Italian Beretta BM-59 folding-stock variant, but required both a modernized and more compact battle rifle based on the original design. Crane developed the Mk14 Mod 0 and the weapon was adopted by the SEALs and was used in combat from 2004. The Mk14 featured a collapsible stock, a skeletonized frame and a shortened barrel, along with a number of Picatinny rails. Interestingly Delta was working on a similar design at the time after producing a field-expedient shortened M14 with the folding stock from an AK in Afghanistan.

The Mk14 proved popular with the SEALs and began to see operational use with Air Force Special Tactics and U.S. Army Special Forces, who appreciated the 7.62x51mm caliber in such a compact package. The Mk14 was further modified with a longer barrel and adopted by the U.S. Army as a DMR, known as the M14 Enhanced Battle Rifle. The USMC also adopted a modified version of the Mk14 as the M39 Enhanced Marksman Rifle.

The battle-proven Leupold 3.5 to 10 power optic was mounted on the Army's M14 EBR, while the Marines used the Schmidt & Bender M8541 3 to 12 power. The SEALs mounted all manner of optics on their Mk14s, including red dot sights like the Aimpoint. Interestingly, the U.S. Army established that with similar optics, the M14 was almost as accurate as the M24 out to 600 meters (656 yards), and gave consistently sub 2 MOA performance.

Mk20 Sniper Support Rifle (SCAR)

Caliber: 7.62x51mm
Unloaded Weight: 4.85kg
Overall Length: 1,080mm
Range: 600m+
Origin: United States

The Mk20 is part of a SCAR family that also includes the Mk16 in 5.56x45mm and the Mk17 in 7.62x51mm. The SCAR system was developed by Belgian firm Fabrique Nationale based upon a U.S. SOCOM requirement for a replacement for a range of weapons, including the M4A1, the Mk11, the Mk12, the Mk14 and the Mk18 CQB carbine. The basic SCAR platform has been made available in both calibers and in a range of barrel lengths from 10 to 20 inches. The platform also features a side-folding stock as standard, apart from the Mk20, which features a fixed, though skeletonized, adjustable stock.

The adoption of the SCAR by U.S. SOCOM was not without incident. The 5.56x45mm variant, the so-called SCAR Light type classified as the Mk16, was combat trialed by the U.S. Army Rangers in Afghanistan in 2009 and returned with ambivalent results versus the Block 2 SOPMOD version of the M4A1. SOCOM eventually announced that it would not be proceeding with the Mk16 order but would continue with adoption of the Mk17 and Mk20, both of which would fill DMR and spotter's rifle roles within the Green Berets, Air Force Special Tactics and Rangers. The Navy SEALs, however, were enamored with the SCAR family and continued with their procurement.

The version of most interest to us is the Mk20 Sniper Support Rifle. Designed to replace the range of 7.62x51mm platforms used by spotters (and snipers) within SOCOM units, it features a full-length Picatinny rail along the top of the weapon to facilitate the mounting of clip-on night sights that work in-line with the weapon's main optics. The weapon's stock features an adjustable cheek piece, and the length of pull can be adjusted for the individual shooter without tools. It is also designed with an adjustable gas regulator to allow the reliable use of a sound suppressor.

Although originally intended to replace the Mk11 and M110 platforms, this has been a slow process, with both SR-25-derived platforms still being heavily fielded in combat by Army SOCOM units. The Mk20, and the entire SCAR family, are superbly accurate weapons and the Mk17 offers a more compact and lighter-weight alternative to the M110. According to operators who have fired both, the M110 maintains a 1 MOA, while the Mk20 can regularly improve on that at sub 1 MOA, almost unheard of accuracy from an off-the-shelf design.

Heckler and Koch PSG-1

Caliber: 7.62x51mm
Unloaded Weight: 7.2kg
Overall Length: 1,230mm
Range: 800m+
Origin: Germany

The PSG-1 was for many years considered the Rolls-Royce of sniper rifles and was employed by the likes of GSG9, GIGN and the SAS. It and the bolt-action Mauser SP66 were developed in the wake of the Munich Massacre as dedicated police rifles for counterterrorist use. Originally only one optic was available, the Hensoldt fixed 6-power scope, although the PSG-1A1 version introduced in 2006 added rail mounts along with a folding stock to reduce the weapon's overall length for transport. Rather unusually, the PSG-1 featured a tripod that acted as a stabilizer and assisted the PSG-1 in attaining its reputation for phenomenal accuracy.

The PSG-1 is still considered one of the most accurate semiautomatic sniper rifles in the world. To be passed from the H&K factory, each PSG-1 needed to be able to fire 50 rounds into a 1 MOA circle at 300 meters, so a grouping of 50 rounds into a target just over 3 inches in diameter. Today it has largely been eclipsed by more recent designs, but it is still in use by the Spanish GOE and Italian GIS national counterterrorism teams.

The PSG-1 attracted poor military sales, as the design was considered too fragile for the rough handling it would receive on the battlefield. The Heckler and Koch MSG-90, MSG-90A1 (modified with a threaded barrel for suppressors) and MSG-90A2 were later introduced as more hardy and lightened versions that were also more keenly priced. H&K also offered sniper versions of their standard battle and assault rifles such as the G3 SG/1.

Heckler and Koch 416

Caliber: 5.56x45mm
Unloaded Weight: 7.2kg
Overall Length: 1,037mm With D20RS Barrel
Range: 400m+
Origin: Germany

The HK416 was developed by Heckler and Koch for and with Delta Force, the U.S. Army's counterterrorism unit. Delta wanted a more reliable 5.56x45mm carbine than their then-issue M4A1 (and originally a side-folding stock, but this fell by the wayside to reemerge years later as part of their Low Visibility Assault Weapon requirement that appears to have been filled by the SIG MCX), particularly during full automatic firing, and the 416 was designed from the ground up with operator input. The first 416 off the production line was put through a 15,000-round torture test, passing without a single stoppage. The rest of that initial production lot went to Iraq with Delta, the ultimate weapon-proving ground.

Although not designed as a sniper weapon, the 416 has been adopted by many SOF units as an intermediate-range sniper platform. The SEALs used the suppressed 14.5-inch barrel version in the famous sniper intervention to rescue Captain Richard Phillips from pirates, and apparently both Delta and SEAL Team 6 employ that version in similar capacities.

Interestingly, the U.S. Marines, including MARSOC, deploy a modified HK416 as the M27 Individual Automatic Rifle (IAR). Equipped with an ACOG optic, the M27 was procured to supplement and largely replace the 5.56x45mm M249 SAW. Marines have found that the weapon can provide either suppressive fire in fully automatic or in semiautomatic mode as a very accurate marksman's rifle. One Marine in Helmand was quoted as saying: "It's very accurate. On single-shot, you can hit 800 yards no problem. I love that you can go from single shot to full auto with the flick of a switch."

HK417

Caliber: 7.62x51mm
Unloaded Weight: 4.2kg
Overall Length: 1,002mm
Range: 600m+
Origin: Germany

The Heckler and Koch 417 was developed in the immediate aftermath of the success of the HK416. A 7.62x51mm version, based on the same design, was almost inevitable and the weapon has been widely adopted as a sniper/spotter's rifle and as a DMR. The HK417 challenges the market position of the SR-25 and similar American designs.

Adopted by the Bundeswehr as the G28, the HK417 has an effective range of between 600 and 800 meters and, unusually for a sniper weapon, it has a fully automatic selector that allows automatic fire at a cyclic rate of some 600 rounds per minute. This makes it particularly of interest as a DMR weapon that may be required to provide suppressive fire or use bursts during house clearing or similar CQB.

The forward hand guard is free-floating to reduce minute vibrations that may affect the accuracy of the shot, and the barrel itself is a 1 in 11-inch twist cold hammer-forged chrome-lined barrel. The weapon feeds from a 20-round magazine and features a number of Picatinny rails to allow optics, lights and lasers to be fitted.

It has seen significant success both within military and police/CT sniping, with the weapon adopted by the British Special Forces Support Group, Germany's KSK and GSG9, Australian SOF and the majority of European counterterrorist units (the 417's civilian version, the MR308, since renamed the MR762, is even in service with Russian SDF in Syria and the Ukraine). Master Sniper Nathan Vinson commented on the Australian issue:

> The SR25s were well worn when we arrived [in Afghanistan] and had been passed on from one contingent to another but were much loved when in service. The HK417 only came into service in late 2011, early 2012 due to the urgent need for a marksman's weapon; it was then also passed onto sniper teams within the battalions. It proved to be a superior weapon on operations and highly regarded by the teams and designated section marksmen.

As this book went to press, another significant adoption of the 417 was announced. A G28 variant of the 417 won the tender for the Compact Semi-Automatic Sniper System (CSASS) programme. A version of the G28E1 is now set to replace all M110s in U.S. Army service, an order eventually totaling over 3,600 rifles.

SVD Dragunov

Caliber: 7.62x54mm
Unloaded Weight: 4.3kg
Overall Length: 1,225mm
Range: 600m+
Origin: Russia

As previously noted, the Russian sniper training between the 1950s and the late 1990s focused on churning out the equivalent of the Western Designated Marksman. Such a marksman was part of every Russian infantry squad, his role to engage targets beyond the 400-meter

(437-yard) effective range of the standard 7.62x39mm AK47 or the later 5.45x45mm AK74. His optics were also valuable in identifying targets for support weapons, like the 7.62x54mm PKM medium machine gun.

It should come as no surprise that the standard-issue Russian sniper rifle of the time, the SVD Dragunov, was much more a marksman's weapon, designed only to engage targets out to 600 meters (656 yards). Apparently it was designed with even closer combat in mind, as it featured a bayonet mount! Despite this anachronism, it also featured backup iron sights should the principal optics fail, a standard today but forward-thinking for its time. The SVD entered service in 1963, replacing the venerable bolt-action Mosin Nagent M91/30 PU, veteran of the Eastern Front.

The SVD typically mounts the PSO-1 scope, which is a fixed 4-power magnification optic with illuminated reticle for shooting in low light. The reticle featured both range-finding marks based on the estimation range-finding method (as described in Chapter 2) and a series of chevrons that theoretically provided hold-off aiming points for targets beyond 1,000 meters (1,093 yards). Although the SVD design may be somewhat dated, the 7.62x54mm round still holds its own as an ideal sniper round with excellent ballistics and lethality.

L129A1

Caliber: 7.62x51mm
Unloaded Weight: 4.5kg
Overall Length: 990mm
Range: 800m+
Origin: United States

Officially known as the Sharpshooter Rifle, the L129A1 was the result of an Urgent Operational Request from British troops in Afghanistan who needed a semiautomatic 7.62x51mm platform to supplement and eventually replace the bolt-action L96A1s and L118A1s they had been pressing into service as a platoon-level Marksman Rifle.

The result was a design from the American firm of Lewis Machine and Tool, which beat competition from the HK417 and the Mk17 SCAR. The L129A1 mounts a Trijicon 6-power optic with bullet-drop compensator and Shield mini red dot sight to allow the weapon to be used at close quarters (the shooter simply ignores the Trijicon scope and uses the red dot affixed to the top of the scope, which is likely zeroed for 50 meters (55 yards) or less).

The L129A1 is fed from 20-round magazines and, unlike its competitors, the HK417 and Mk17, the L129A1 is semiautomatic only (although the L129A2 version is selective fire). Rated at sub 1 MOA in accuracy, the L129A1 fires a 155-grain match-ball round that has resulted in hits at 1,000 meters (1,094 yards) in Helmand. Indeed, the weapon proved so popular with British troops in Afghanistan that snipers have used it as a primary sniping weapon when the extended range capability of the L115A3 was not required.

ANTIMATERIEL RIFLES

The U.S. Army describes these impressive weapons thus:

> These heavy sniper rifles were originally intended as antimateriel weapons for stand-off attack against high-value targets, such as radar control vans, missiles, parked aircraft, and bulk fuel and ammunition storage sites. It is their ability to shoot through all but the heaviest shielding material, and their devastating effects, that make them valuable psychological weapons. The ability to shoot through common urban building materials makes these large weapons valuable as countersniper tools.

The antimateriel rifle was, as the U.S. Army describes, first intended as a long-range rifle to engage enemy equipment and vehicles, not enemy soldiers. Its early adoption by SOF units saw the nascent possibilities of its deployment against human targets: leaders, radio operators and of course, enemy snipers. This owed more to the tremendous stand-off capability provided by the rifles rather than the inherently deadly .50BMG round, more commonly found in the Browning M2 heavy machine gun. The counterinsurgencies in Afghanistan and Iraq proved the validity of the antimateriel rifle as an antipersonnel weapon.

Like most firearms, and particularly those used by snipers, there are a number of myths that surround the antimateriel rifles and the .50BMG round itself. Chief among these is the often-held contention that it is illegal under the Geneva Conventions to employ the .50BMG against human targets. Exactly where this tall tale emerged from is difficult to identify, although it may be related to a Hague Conventions edict against the use of exploding bullets. Either way, there is no restriction in the Rules of Land Warfare or anywhere else on the use of the .50BMG and similar calibers on human targets.

The other commonly reported myth is that even a miss by a .50BMG will cause injury. Where this originated is anyone's guess, but it was a common myth within the U.S. military for generations. As even a passing understanding of ballistics would show, a bullet must of course hit a target to cause physical damage. Granted, with explosive weapons, a near miss might cause concussion or fragmentary wounds (and the explosive-tipped Raufoss Mk211 might be the round that actually gives some credence to the myth after all), but otherwise a hit is still required.

Barrett M82A1 (M107, M107A1, M82A3 G82)

Caliber: .50BMG
Unloaded Weight: 16kg
Overall Length: 1,447.8mm
Range: 2,000m+
Origin: United States

Ronnie Barrett developed the rifle that bears his name back in the early 1980s. The first examples were produced in 1982, leading to the rifle's name, the M82. An improved variant with military input emerged in 1986, the M82A1. It was first purchased in small numbers by the U.S. Navy SEALs and the U.S. Army Special Forces before it was adopted by both the U.S. Army and Marine Corps immediately prior to Operation Desert Storm, although Barrett's first military sales were to the Swedish Army. There were a number of other .50BMG designs that also saw limited adoption, including the Haskins and the McMillan (that later became the Mk15).

The Barrett, however, was finally officially adopted as the M107 in 2002. Originally designated the SASR for Special Application Scoped Rifle, the M107 is today designated by the U.S. Army as the Long Range Sniper Rifle or LRSR. The U.S. Marines have adopted a version with a full-length Picatinny rail along the top of the weapon and a rear monopod that is termed the M82A3, topped with a Schmidt & Bender optic. The Army M107 mounts another variant of the Leupold Mark 4 common to all U.S. Army standard sniper weapons, this time in the variable 4.5- to 14-power configuration.

The M82A1 platform is widely used by most NATO nations, including adoption by the Bundeswehr as the G82. Most SOF units maintain a number of M82A1s. Nathan Vinson mentioned that the Australian SOF had access to both the bolt-action AW50 and the Barrett but said: "SOCOMD [Special Operations Command] canned the AW50 fairly early in the piece, mainly due to its weight and size. As you may know, the Barrett could be broken down, and also being semiauto it met their needs greater than the AW50."

The weapon does have its own disadvantages. It is not considered one of the most accurate platforms. Staff Sergeant Dillard Johnson commented on the Barrett that he used in Iraq: "The Barrett isn't as accurate as most specialized sniper rifles, but then again most of the engagements we had would be considered short range for a sniper, since we worked in urban areas."[6] Although the Mk211 and similar match ammunition can improve the weapon's accuracy, most shoot between 2 and 3 MOA. Nor is it lightweight, at almost 16 kilograms loaded with a ten-round magazine.

It is also punishing on the shooter. The SASR/LRSR could never be considered a particularly pleasant weapon to fire, and Chris Kyle even said this about the rifle:

> The fifty is huge, extremely heavy, and I just don't like it. I never used one in Iraq. There's a certain amount of hype and even romance for these weapons, which shoot a 12.7x99mm round. There are a few different specific rifles and variations in service with the U.S. military and other armies around the world. You've probably heard of the Barrett M82 or the M107, developed by Barrett Firearms Manufacturing. They have enormous ranges and in the right application are certainly good weapons. I just didn't like them all that much.
>
> Everyone says that the .50 is a perfect anti-vehicle gun. But the truth is that if you shoot the .50 through a vehicle's engine block, you're not actually going to stop the vehicle. Not right away. The fluids will leak out and eventually it will stop moving. But it's not instant by any means. A .338 or even a .300 will do the same thing. No, the best way to stop a vehicle is to shoot the driver. And that you can do with a number of weapons.[7]

The SASR/LRSR has a rated effective range of some 2,000 meters. Again, as with most sniper platforms in combat use, the platform has successfully engaged targets at much greater distances. The rifle still holds the record, at the time of writing, for the longest-range kill at 2,815 meters (3,079 yards) by members of Australia's 2 Commando Regiment in Helmand Province in 2012, as detailed in an earlier chapter. At the reported range, the bullet drop would have been approximately 300 feet, while the .50BMG would have lost enough velocity to have effectively become subsonic.

The Barrett is certainly capable of some astounding shooting. In at least one incident in Iraq, an insurgent was engaged and shot and killed through a parked car with the .50BMG M107. According to Chris

Martin, the author of *Modern American Snipers*, a Delta sniper made a shot in the mountains of eastern Afghanistan in 2001 at an impressive 2,500 yards (2,286 meters).

In a post-script to the Barrett story, the rifle was named the official state firearm of Tennessee in February 2016, honoring Ronnie Barrett, who developed the rifle, and who has been a lifelong resident of the state. Previously similar official state firearms have been lever-action carbines and black powder rifles. The appointment of the Barrett by Tennessee brings the unique distinction into the 20th and now 21st century. The success of the Barrett sees no sign of waning.

Mk15 Mod 0

Caliber: .50BMG
Unloaded Weight: 12.2kg
Overall Length: 1,447.8mm
Range: 2,000m+
Origin: United States

The Mk15 Mod 0 is the standard .50BMG platform issued within U.S. Special Operations Command (SOCOM) units. It is a version of the McMillan TAC-50 that was so successfully employed by Canadian snipers during Operation Anaconda in 2002 and the McMillan M88 that was used by Navy SEALs for much of the 1990s. Firing from a bolt action, the Mk15 is considered to be a more accurate antimateriel rifle than the more popular Barrett designs.

The Mk15 has seen extensive service with U.S. Navy EOD teams in Iraq and Afghanistan, who use the explosive Mk211 to destroy IEDs at safe stand-off ranges. Additionally, it has been heavily employed by Navy SEAL teams in both nations, although the weapon's size and bulk negates against its everyday use. Most SEAL snipers carried instead the .300 Mk13 or .338 McMillan TAC-338, for instance.

Accuracy International AW50F

Caliber: .50BMG
Unloaded Weight: 15kg
Overall Length: 1,420mm
Range: 2,000m+
Origin: United Kingdom

The British Army and Royal Marines adopted the AW50F as the L121A1, fitted with the Schmidt & Bender 3-12 variable-power magnification PM II scope and side-folding stock. The AW50F, and fixed-stock variant the AW50, were designed at the turn of the century based on the successful Accuracy International AWM platform. The British adopted the weapon primarily as an EOD tool for long-range destruction of IEDs and unexploded ordnance (UXOs).

United Kingdom Special Forces also procured the weapon, where its increased accuracy over the Barrett was preferred. The Australian Army also obtained a small number of AW50Fs for both SOF and EOD roles. Chris Kyle was quoted as rating the Accuracy International AW50 as his preferred choice for a .50BMG antimateriel rifle: "The one .50 I do like is the Accuracy International model, which has a more compact, collapsible stock and a little more accuracy; it wasn't available to us at the time."

SPECIAL PURPOSE RIFLES

Small-caliber rifles have long held a specialist niche within military and police sniping. The humble .22LR and .22-250 both have seen use. Semiautomatic .22LR Ruger 10/22s are used by Israeli Defense Force snipers as a less lethal option to be employed against rioters throwing petrol bombs. "Less lethal" options are designed to be more likely to wound rather than kill in comparison to standard bullets and typically include such weapons as baton rounds ("rubber bullets"). Known as the Toto within the IDF, the .22 Long Rifle round is now also used by

Israeli police sniper teams to shoot rock-throwing demonstrators as a less lethal option than using a larger-caliber round like the 5.56x45mm.

Incredibly, the British SAS also originally recommended the .22LR caliber as a viable option in a report prepared by then Captain Andy Massey, written just two weeks after the Munich hostage taking. Massey wrote:

> German police discarded a valuable option by their refusal to consider a swift, determined, close assault as a means of "springing" the prisoners. This left them with one alternative, namely, at some stage in the proceedings to get involved in a fire-fight from a distance. The wrong weapon was used by the police. A more suitable weapon would have been a silenced .22 specialist sniping rifle. This would have helped achieve surprise.

It must be remembered that at the time there was little in the way of research into subsonic loads for the 5.56x45mm and 7.62x51mm, let alone modern suppressors. The suppressed .22LR with subsonic loads is an incredibly quiet combination, despite its less than stellar wound ballistics. Remember also that the SAS and SBS counterterrorist team deployed for many years with both suppressed and non-suppressed Tikka .22-250 caliber rifles, a higher-velocity cousin of the .22LR.

European counterterrorist units have also long maintained bolt-action .22LR rifles with integral suppressors for close-range, near-silent shooting and for shooting out street lights. Renowned Finnish gunmaker Sako, for instance, produced the integrally suppressed .22LR Mark 3 SSR (Silenced Sniper Rifle) for many years for a number of European teams.

Along with integrally suppressed .22LR and .22-250 rifles, many weapons have been built from the ground up as suppressed-sniper rifles. The French PGM Commando 7.62x51mm integrally suppressed rifle is but one such example. The PGM Commando with side-folding stock was apparently designed expressly with covert operations in mind, in much the same way as the Accuracy International AWS we have earlier discussed.

The Russians too have developed a number of suppressed-sniper systems, the 9x39mm Vintorez VSS being a case in point. The 9x39mm round was itself designed to be used with a suppressor, being a heavyweight subsonic load. Knight's .30 Silent Sniper Revolver that will be covered in some detail in the interview with John McPhee in Appendix 1 is another example of a sniping platform built from the ground up for silence.

At the other end of the spectrum are the special-purpose rifles built to fire explosive rounds far more usually encountered in cannons. Barrett's XM109 fires a high-explosive 25x59mm round from a five-round detachable box magazine. Its intended purpose is more for stand-off destruction of IEDs and mines, although it could certainly be used as a sniper rifle and will penetrate the armor of most Russian-designed armored personnel carriers. The Austrian firm of Steyr also developed a remarkable heavy-caliber precision rifle in the form of the IWS 2000 in 15.2mm. It fires an APFSDS (Armor-Piercing Fin-Stabilized Discarding Sabot) smoothbore 308-grain tungsten dart that will reportedly penetrate 40mm of armor plate at 1,000 meters (1,093 yards).

PERSONAL PROTECTION AND SPOTTERS' WEAPONS

Nearly every army, and certainly every police and counterterrorist unit, issue their snipers with an automatic pistol as a sidearm. As we have previously noted, thanks to combat experience in Iraq, many forces also issue carbines to their snipers, or at the very least to spotters.

U.S. Army sniper teams are each issued a 5.56x45mm M4A1 carbine, with one mounting the 40mm M203 grenade launcher. They are also issued two 9x19mm M9 pistols. In cavalry sniper teams, the team leader and spotter both carry M4A1s, with the spotter also carrying the M203 or M320. Their M4A1s are fitted with the excellent 4-power Trijicon ACOG optic.

German spotters would carry the 4.6mm Heckler and Koch MP7, although they often switched out for a longer-range 5.56x45mm G36 in Afghanistan. From 2007, every British soldier serving in Afghanistan, including snipers and platoon marksmen, received a 9x19mm SIG pistol or 9x19mm Browning Hi-Power for personal protection. British sniper James Cartwright explained the requirement:

> If, for example, we went into the Green Zone and encountered a group of Taliban with AK47s, at close range, we would have been in serious trouble as we only had bolt-action .338 sniper rifles that only hold five rounds in the magazine. We were eventually successful, our argument being that if we were moving through alleyways or around buildings, we needed to have our pistols drawn so that we could react quickly and effectively to a threat at close quarters.[8]

At least two British snipers had cause to use them in close-range encounters with insurgents while fighting in and around Afghan compounds. Colonel Richard Kemp recounts one such encounter:

> Suddenly a fighter popped up from behind the wall with an RPG, right beside Bailey [the sniper]. Bailey smacked him straight in the face with his pistol, then, moving back, fired a full magazine of 9mm bullets into him as he collapsed, [and] dropping into the cornfield ... Bailey couldn't believe what had happened. This was the second time since arriving in Helmand that he, a sniper, had ended up killing a Taliban fighter at close range with his pistol.[9]

Another literally ran into an insurgent as he moved down an alleyway. He managed to draw his 9x19mm Browning Hi-Power and fire off a magazine, while the insurgent fired his AK47 at point-blank range. The sniper hit the insurgent with at least three rounds and managed to pull back, throwing a fragmentation grenade around the corner to dissuade any pursuit from other insurgents. Later the sniper discovered a notch gouged along the slide of his Browning: evidently one of the insurgent's

AK47 rounds came dangerously close to hitting him. He was otherwise unhurt.

The British also issue each sniper with his own L85A2 assault rifle. When patrolling with an infantry unit, the snipers would carry their sniper rifles in a drag bag across their back and patrol with the L85A2 as their primary weapon. This accomplished two things: firstly, it wasn't readily apparent they were snipers, a priority target for the insurgents, and secondly they had a compact weapon that could be used to rapidly engage the opposition. Estonian sniper teams that served alongside the British in Helmand didn't bother too much with appearances and carried far more exotic wares, such as suppressed 5.56x45mm Galil SAR carbines and suppressed 9x19mm Heckler and Koch USP pistols.

SOF units would typically carry whatever they liked. SEAL sniper elements would see the snipers also carrying their own 5.56x45mm M4A1s or Mk18s (an even shorter version of the carbine) to supplement their primary weapon. Delta snipers in Iraq tended to deploy with the semiautomatic 7.62x51mm SR-25K and carried pistols as backup (some also carried a 40mm standalone grenade launcher). For SOF sniper teams in particular, the strong chance of being surrounded and cut off from immediate assistance meant that they were forced to carry enough firepower to extricate themselves, should the worst case scenario occur.

Afghan and Chechen War historian Les Grau adds a counterbalance and explains the standard loadout of a Russian sniper in Chechnya:

The sniper carries his sniper rifle as well as an assault rifle or machine pistol for close-in fighting. He also carries a night-observation device, dry rations, hard candy, a flare pistol with a red flare, a grenade, two shelter halves, and a shovel. Sometimes he also carries a radio. In the mountains, he carries a ski pole to help him climb. He wears a mask to hide his skin tone.[10]

As noted earlier, the Russian sniper is trained to use the flare to bring in an artillery or mortar strike on his own position in case he is compromised and overrun. The single fragmentation grenade carried serves a similar purpose: up against the Chechens and knowing their reputation for barbarity when it comes to prisoners, the sniper will use the grenade to kill himself and as many of his attackers as possible. While such tactics may seem extreme from a Western approach, the beheadings and burnings of prisoners captured by Islamic State suggest that such precautions may become more necessary for Western SOF snipers operating in Iraq and Syria. As this book was going to press, Russian media reports indicated that exactly this scenario had occurred in Syria during operations to retake Palmyra. A Russian special operator became separated and surrounded by Islamic State terrorists and, instead of allowing himself to be captured, ordered an emergency airstrike on his own position, killing himself in the process.

A Spetsnaz sniper explained such circumstances earlier in the Soviet-Afghan War.

Being captured by the Mujahideen was one thing you did not want to do. It was always in the back of your mind that the last grenade was yours. What they done there was skin people alive, they ah, I guess desecrated bodies, cut off heads, awful shit. It was some medieval stuff. Some of the bodies that we came across ... had all kinds of traces of extreme torture.

So one thing you didn't want to do was be captured, that's for sure. There was several instances where troops called fire in upon themselves. It's a common thing in guerrilla warfare for the enemy to hug your position thinking that your artillery will not fire on them or if they do fire they will be ineffective, like within 100 meters or so. So in that case if you're running out of ammo and are receiving fire from all sides you might want to draw them in a little closer and call fire in upon yourself. There were instances like that.

MODERN SNIPER EQUIPMENT

Laser Rangefinders

Laser rangefinders have revolutionized sniping. The first examples were designed for use by forward observers and were both heavy and somewhat delicate. Most military versions now resemble armored binoculars. The SOFLAM or Special Operations Forces Laser Aiming Module has also been used to lase targets for snipers versus its traditional role of illuminating targets for laser-guided aerial bombs.

The most common type in use by snipers today is the Vector 21. This device is accurate out to 4,000 meters within 3 meters. It can also accommodate angles up to 35 degrees to assist in high- or low-angle shooting. Finally, it incorporates a GPS receiver to plot targets. The Bushnell Elite 1500 laser rangefinder is also widely used, particularly by police snipers. This compact device provides a range with accuracy within plus or minus a meter out to 1,500 meters. British troops commonly use the PLRF10 Pocket Laser Rangefinder that can range targets out to 2,500 meters at a 6-power magnification.

Along with laser rangefinders, another now common device that emerged in the 1990s has improved the accuracy of generations of snipers: the electronic wind gauge. Although such devices provide an accurate read of the wind, it only registers the wind at the location of the device.

Night Vision

The biggest recent advance has been both the reduction in size of night-vision devices for the sniper and the ability to either use the one optic for both day and night shooting or mount a device in-line with the sniper's existing optic, allowing them to choose when to add the night-vision capability. Almost as important has been an increase in battery life for many devices.

Military night-vision devices work in two ways, infrared and thermal. Infrared can be either passive or active. Passive infrared devices magnify all available natural light to generate an image through the viewer. Active uses an infrared light source, such as that projected by infrared illuminators that can be weapon-mounted, like the AN/PEQ-14, or mounted on a vehicle or aircraft. The AC-130 Spectre has such a system, as do various models of UAV like the Reaper: these are more like infrared searchlights that can be used to illuminate entire areas to soldiers wearing night vision.

Most snipers will mount some type of infrared laser to their secondary weapon if not their primary (many SOF snipers using semiautomatic platforms will routinely mount such a device, as they will often be shooting at relatively close range during darkness hours). This projects a laser beam that is only visible through night-vision devices like goggles and can be used to mark or indicate a target.

Thermal imaging, on the other hand, works by displaying the relative heat signatures of objects. They have an advantage when engaging targets in difficult, concealing terrain and can see through smoke and fog. They work best at night, as the contrast is obviously far more pronounced, but can be used during the daytime, particularly in urban environments.

U.S. snipers typically mount the AN/PVS-29 or AN/PVS-30 CSNS (Clip-on Sniper Night Sight) in front of their Leupold optics. The CSNS can be easily attached or detached, thanks to its Picatinny rail mount. The older AN/PVS-22 Universal Night Vision Adapter worked on a similar principle and was often deployed with the Mk11 and M110. They also have available the AN/PVS-10, which is both a day and night optic that offered a fixed 8.5-power magnification and was used extensively during the wars in both Iraq and Afghanistan.

The AN/PVS-13 Thermal Weapon Sight, which was often employed as a standalone thermal imager by spotters, is now issued as the improved AN/PAS-13G Lightweight Thermal Sight and can be used with both magnified optics like the standard Leupold Mk4 and combat optics like the ACOG. The British use a special Sniper Thermal Imaging Capability (STIC) device that likewise mounts via Picatinny rail directly ahead of the Schmidt & Bender optic on their L115A3s.

They also have available the AN/PVS-27 night-vision scope, which can be mounted forward of the optics on their L129A1 Marksman Rifles in the manner of the CSMS. British spotters carry the VIPIR-2S+ handheld thermal imager, as do a number of European nations.

Snipers will of course also use their personal night-vision devices, such as the AN/PVS-14 and AN/PVS-16 or the latest generation AN/PSQ-20 that combines night vision with image intensification. These are all mounted on the front of their helmets and can be swung up and out of the way when using the scope on the sniper rifle, although snipers prefer a weapon-mounted night-vision device, as attaining a sight picture through the scope while wearing helmet-mounted goggles can be difficult.

Spotting Scopes

Spotting scopes have also advanced far from the commercial hunting models common during the latter half of the 20th century. The standard-issue American AN/PED-3A Target Locator Module carried today looks like an oversized pair of armored binoculars incorporating a laser rangefinder and 7-power optics.

The spotting scope of today's spotter or observer is normally several times more powerful than the sniper's actual optic. Many police departments use the 40-power tripod-mounted Burris spotting scope, while the M151 Enhanced Spotting Scope used by U.S. Army snipers, a militarized version of the Leupold Mark 4, features a mildot reticle to allow for easy adjustment calls and an impressive 40-power magnification. The British military use the similar Leupold Golden Ring Mark 4 spotting scope with variable 12- to 40-power magnification.

The next generation of optic will be something like the L-3 IBRS or Integrated Ballistic Reticle System that can be used as a 4- to 22-power sniper optic or as a standalone spotting scope. Incredibly, the IBRS offers a laser rangefinder, ballistic computer with weather sensors to measure humidity, density and temperature, a GPS compass and sensors that take into account the angle the scope is being used at, and automatically modifies the shooting solution to take all of these disparate elements into account. Put simply, the sniper or spotter uses the IBRS to lase the target and the system does the rest.

Field periscopes dating back to World War II were brought back into service by Russian snipers during the Chechen wars, as they enabled the sniper to observe potential targets without exposing their head. The Micro Times SSVZ-1 Handheld Periscope takes the concept and updates it. Similar to the periscopes used by Russian and German snipers, the SSVZ-1 is far more powerful, with a variable 4- to 9-power optic, and far less obvious to anyone but a trained observer. It also telescopes down to make a handy package to store in an assault pack or day sack, and has proven popular with Coalition snipers in Iraq and Afghanistan.

Suppressors

The U.S. Army's SOTIC sniper manual is clear on the importance of the sound suppressor:

The suppressor is a device designed to deceive observers forward of the sniper as to the exact location of the weapon and the sniper. It accomplishes this by disguising the signature in two ways. First, it reduces the muzzle blast to such an extent that it becomes inaudible a short distance from the weapon. This makes the exact sound location extremely difficult, if not impossible, to locate. Secondly, it suppresses the muzzle flash at night, making visual location equally difficult. This is critical during night operations.

When a rifle, or any high muzzle velocity weapon is fired, the resulting noise is produced by two separate sources. Depending on distance and direction from the weapon, the two noises may sound as one or as two closely spaced different sounds. These sounds are the muzzle blast and the ballistic crack, or sonic boom, produced by the bullet.

Master Sniper Nathan Vinson argues that suppressors are a key part of the modern sniper's core equipment. "Absolutely essential, helps mask the shot and the sniper position … a must."

The use of suppressors will, however, affect the point-of-aim of sniper rifles. Many counterterrorist teams, including the British SAS, maintained both a suppressed and non-suppressed model of their sniper rifles, as each would have a slightly different zero.

Certain modern suppressors such as the Surefire SOCOM virtually eliminate this, allowing a weapon to be zeroed without a suppressor attached but later deployed with the suppressor attached with minimal to nil change in the point of impact. The Surefire SOCOM has another key advantage over many of its competitors: it can be mounted on a range of platforms. For instance, the SOCOM556 can be attached to an M4A1, the HK415A5 or the Mk12 without requiring a separate suppressor for each.

In Helmand, British snipers often used suppressors on their 7.62x51mm L96A1s and .338 Lapua L115A3s. Author and journalist Toby Harnden explained the advantages, using the example of a Welsh Guards sniper pair who racked up 75 kills in 40 days using suppressors:

> Although a ballistic crack could be heard, it was almost impossible to work out where the shot was coming from. With the bullet travelling at three times the speed of sound, a victim was unlikely to hear anything before he died.
>
> Walkie-talkie messages revealed that the Taliban thought they were being hit from helicopters. The extraordinary thing was that it took the Taliban a month to begin to realize what was killing so many of them. Up until then, fighters kept queuing up at the same firing points only to be killed one by one.[11]

Ballistic Computers and iPhone Apps

"There's an app for that," goes the common advertising refrain, and, perhaps incredibly, there are any number of apps for sniping. Some are more suited for the casual civilian shooter, but some have been developed with military input and are used by sniper teams deployed operationally. The advantage of a ballistic computer or app is that a firing solution can be developed faster and more accurately than even the best trained and most experienced sniper. Nathan Vinson explained:

> We have been playing with various systems, including iPhone downloads, for close to six or seven years now. I think you will find that all teams now have a variant of one kind or another. We allowed them on the Team Leader's course but only after they had successfully passed the marksmanship package and only on the final mission phase of the course.

Clothing

Even the Ghillie has seen technological improvements. The latest U.S. version, the GSAK or Ghillie Suit Accessory Kit, is both flame-retardant and counter-thermal imaging, meaning that the suit helps dissipate the thermal profile of a sniper in an enemy thermal viewer. The next evolution of sniper camouflage will be Nemesis battledress uniforms that effectively shield the snipers from easy detection. Active camouflage is also being researched that would see the camouflage pattern of the uniform and the Ghillie automatically change to match the surrounding environment.

Infrared heat source-reducing coatings can also be applied to the sniper rifle itself. Matched with the Nemesis reduced-infrared-signature uniforms and Ghillie suits, such options are being adopted in an effort to counter relatively inexpensive infrared optics that insurgents and terrorists may begin to field in an effort to identify sniper hides.

Snipers also tend to wear either plate carriers or even concealed light body armor, like the Ultra-Low-Visible Concealable Body Armor designed to be worn under a shirt yet retaining the ability to defeat the most common threat, the Soviet 7.62x39mm round. Plate carriers evolved from SOF requirements for lighter-weight body armor that was less restrictive to movement and more comfortable when deployed in the prone position. They feature front and back trauma plates that will stop rifle rounds, but typically feature less protection than the standard-issue Interceptor or Osprey combat body armor.

Guided Rounds, Tracking Point and the Future

With the advent of optics like the L-3 IBRS and ballistic apps that will provide an accurate firing solution for today's sniper, the next step in terms of technology is the guided round. This may take one of several forms. The best known are the 25mm guided projectiles developed for the XM-25 Punisher that have seen field combat trials in Afghanistan to generally positive reception. The XM-25 allows its rounds to be set to airburst over the enemy should they be behind cover, or indeed dual-purpose rounds can be programmed to punch through a wall before exploding.

So-called smart sniper rounds are also currently under development at Sandia National Laboratories. The Extreme Accuracy Tasked Ordnance is surprisingly fired from a smoothbore weapon (like something of a high-tech shotgun). The reason for the smoothbore is because of the projectile, which has fins like a fin-stabilized sabot cannon round.

Behind an optical sensor in the nose sits a tiny computer processor that reacts to changes in direction fed from that optical sensor. The sniper, or the spotter, directs the Extreme Accuracy Tasked Ordnance via an infrared laser illuminator in much the same way as a SOFLAM designator directed aerial bombs. While the target is painted by the laser, the round will follow that laser, even if the target moves or attempts to hide behind cover.

Another experimental round is the EXACTO being developed by the Defense Advanced Research Projects Agency or DARPA. The EXACTO is fired from a rifled bore and can be used from a .50BMG antimateriel rifle. While the details are rightly classified, the EXACTO appears to be guided more like an antitank guided weapon with changes to the flight path of the bullet made through a scope.

No mention of future sniping technologies would be complete without discussing TrackingPoint. The system made a big splash in the media as it was trumpeted as being able to turn anyone into a sniper. The reality of course is a little different. TrackingPoint is a system that effectively develops a firing solution for the shooter, one that will change to take into account the movements and actions of the target. The shooter simply uses a TrackingPoint-enabled rifle optic to acquire their target. Simply put, they lase the target to lock it in and the system largely does the rest. Only the variables of wind speed and direction both at the target and during the bullet's flight path are not yet automatically calculated. The rifle continues to track the target and will fire once the optimum firing solution presents itself.

The downside to technology such as this is that it is always vulnerable to enemy action. TrackingPoint itself was allegedly the target of hackers, who claimed to be able to take control of the rifles remotely in much the same way as the Iranians have claimed to have hacked U.S. drones. In fact, should such guided technology become commonplace on the battlefield, so would the countermeasures, and in today's world of advanced technology, how long would it be before an app was available to counter these guided projectiles?

APPENDIX 1

AN INTERVIEW WITH A SPECIAL OPERATIONS SNIPER

John McPhee is a former Army Ranger, Green Beret and veteran of Delta's A Squadron, including its recce sniper troop. He is perhaps best known by his Delta callsign, "Shrek." In 2001 McPhee hunted Usama bin Laden in the mountains of Tora Bora, was present when Saddam Hussein was captured and was instrumental in the hunt for the Jordanian leader of al Qaeda in Iraq, Musab al Zarqawi.

In 2002, he was famously sent out on a highly dangerous singleton reconnaissance mission against an Afghan who had assisted bin Laden in his escape from the mountains and caves of Tora Bora. To blend in with his surroundings, McPhee donned an Afghan *shalwar kameez* (the traditional two-part Afghan shirt and trousers) and traded his much-loved customized 7.62x51mm Heckler and Koch G3A4 for a folding-stock AK.

The operation was not unusual for the unit's recce snipers in Afghanistan. McPhee conducted covert surveillance of the target building after infiltrating hidden among Afghan civilians on a bus. The footage he captured enabled a Delta assault force to breach the compound and capture the targeted individual. The mission was a good example of the types of operations Tier One snipers were and are being called upon to accomplish, even in an age of micro UAVs and other ISR assets.

He later became command sergeant major at Delta and completed eight combat tours. McPhee retired in 2011 after over 20 years serving within U.S. special operations, ten of those in near continuous combat, from the Balkans and Central America to Afghanistan and Iraq. Today McPhee runs training programs in small arms, including pistol and carbine courses across the United States and internationally. His website is www.sobtactical.com.

The author had the unique opportunity to speak with this former Delta recce sniper about his favored rifles and equipment and his thoughts on military sniping. Firstly I asked his opinion on the age-old bolt-action versus semiautomatic sniper rifle debate.

I'm a semiauto guy. Here is the deal in the real world of combat: there is not that much difference in accuracy. Meaning there is no such thing as perfect, so it's hard to judge accuracy. Also on the battlefield you only need about 2 MOA out to about 800 [yards] to get a kill shot on a chest and 500 [yards] for a headshot. So all this extreme accuracy of a bolt gun is not needed.

The big downside to bolt guns is the time [delay] between rounds. I know guys say they are just as fast with a bolt gun, but it's just not realistic. Semi has a huge advantage here. When I was a young Ranger we had Rem 700 [the bolt action M24] and it had a 5-round magazine. Do you want to go to war with 5 rounds inside the gun and have to chamber every round? Last, it's easier to stay inside the scope with a semi because your hands stay on the gun the entire time.

In terms of the ideal military sniper caliber, he holds similarly strong views.

I would say .338 or .50. Here is the deal: most snipers are undertrained, so you need a caliber that anyone can get behind the gun and shoot. I teach several sniper classes per year and what I see is most guys don't have the fundamentals to shoot a .308 (7.62x51mm) or .300 out to 1,000 [yards]. Shooting is like an ice cream cone. The tip is small and the farther you get away from it, the bigger it gets on the ice cream end. Same as a bullet: the farther the bullet gets away from the barrel, the bigger any mistake becomes. What might be a ¼ inch at 100 [yards] could turn out to be a miss at 1,000 [yards]. Most guys lack the basic skills to get even a solid group at 1,000 [yards].

He still sees a place for the .50BMG within military sniping, but feels its weight and size negate its regular use by dismounted sniper teams.

(The) Barrett .50 is great for vehicle-based shooting. Mount it in your vehicle gun turret and now you have a long-range accurate gun. There have been many times where this was better than a M2 [.50BMG heavy machine gun]. However, the Barrett is not so good for traditional sniping. This also goes for a .338 semiauto. They are very bulky and heavy to carry. These guns were made for bench shots, not being man portable.

McPhee was instrumental in the development of the shortened version of the Knight's Armament Company's 7.62x51mm SR-25, known informally as the SR-25K, after experiences in Afghanistan, so it was hardly surprising that the weapon featured among his favored rifles, the others being the 50BMG McMillan TAC-50, the Knight's .30 Silent Revolver Rifle and the Accuracy International .300 and .338 platforms.

The unusual Silent Revolver Rifle or SRR from KAC deserves special mention, due to its rarity and method of operation. McPhee is still enthusiastic about the weapon. "I was the last one to give mine up in the unit. It was an awesome suppressed platform. Anything inside of 200 meters I owned! Loved that gun." The SRR was based on a rechambered Ruger Super Redhawk revolver and was fitted with an integral suppressor.

The rifle fired a .30-caliber round that was fitted into a custom-made sleeve that sealed the gap between the frame and cylinder, allowing a revolver-based design to be effectively suppressed. Obviously, being based on a revolver, the weapon only held six rounds and was intended as a close-range suppressed sniper rifle to be used on deniable missions (hence the revolver design that didn't eject the spent brass).

The sound signature of the Silent Revolver Rifle was only a reported 119db. The sound level of the weapon's hammer dropping on an empty chamber was 112db, indicating just how extremely quiet the SRR was. The suppressor also eliminated any visible muzzle flash. In terms of accuracy, the SRR also delivered. 2 MOA groups at 100 meters were apparently consistently recorded. McPhee even fired it over the head of former Defense Secretary Donald Rumsfeld at a demonstration of sniper-initiated assault capabilities at Delta's compound.

I initiated the assault, but shooting all the targets as a sniper. The assaulters would hear my shots and then commence the assault with explosives. I would be up on a telephone pole in blue jeans and a blue shirt, looking like an engineer or electrician. People thought I was fixing the electrical pole; I would have the rifle slung on my chest so you couldn't see it from where you were standing. I would pull my pants down so you could see my butt crack so I looked like a legitimate electrician! After the assault, when I shook hands with Donald Rumsfeld, he looked at the gun and was like "who are you?"

I told him I was the sniper that initiated the assault by shooting over his head! He was so amazed by me and the gun he didn't want to let go shaking hands. He couldn't believe that I was a sniper, not an engineer [and] he couldn't believe the weapon I had used to shoot over his head. All the commanders were like "sir, we got to go," because I'm not the guy they want him talking to!

He also carried a range of 7.62x51mm Heckler and Koch platforms over the years, including a customized HK21E light machine gun that he fitted with a 4-power ACOG during the hunt for bin Laden in Tora Bora. Several of his fellow recce snipers on the mission carried versions of the Knight's SR-25. He preferred the veteran German design for two reasons: "1) Kill shot out to 800 yards only required 2 MOA. 2) The HK21 not only [does] 2 MOA, it does single shot and can do belt feed, a much more versatile weapon for the battlefield."

The HK21E was superbly accurate for a weapon of its type, chiefly due to the closed-bolt design and, as McPhee notes, the unusual semiautomatic selector setting, uncommon on machine guns of any type. He felt it could hit precision targets as well as the SR-25, but also offered a suppressive fire capability should he need it.

He believes the biggest advance in sniper technology in recent years has been the gridded Horus reticle and ballistic aps.

The biggest advantage has been Horus style reticles and ballistic computers. In the old days, we had a swag [Scientific Wild Ass Guess] formula for everything. What we were good at was judging distance, calling winds, and milling. What we did nothing about was scientific data like weather or air density. Now every smart phone can take the scientific and use it to make better shots.

The Horus is a non-caliber-specific scope reticle that drastically speeds up making shot corrections. "Horus Reticles are best for any distance shooting. Also a reticle that is higher than the center of the scope gives you more ballistic drop. More drop you can have, the farther you can shoot," McPhee added. He feels that guided systems like TrackingPoint will be the "way of the future, however TrackingPoint is not it."

He is blunt in his estimation of insurgent snipers, saying he'd never encountered an actual sniper: "they had sniper weapons but sucked at sniping skills." He is also dismissive of the legend of the Afghan mujahideen sniper. "Muj marksman? Hahahahahahaha. More like the CIA making the muj successful against the Russians." When asked what weapons he'd seen insurgent snipers use, McPhee said that "we always see the same things: Dragunovs."

On the impact of the war on terror on sniping schools, he agrees with the U.S. Marine Scout Sniper Course's recent changes to place greater emphasis on marksmanship rather than fieldcraft. "[It's] smart: if you can't hit the enemy, being sneaky or good in the woods don't really matter." He also feels that the Designated Marksman concept perhaps missed one key point. "Every soldier should be a marksman. War would be easier if everyone killed who they aimed at. Marksmanship is not stressed enough."

On specific techniques, McPhee explained he held little of the reluctance to shoot through glass that many of his contemporaries hold, likely due to the extensive training and experience in shooting in support of hostage rescues and personnel recoveries that the recce snipers receive. "Take the shot and shoot straight away. There is a lot of bullshit about this. I've never seen a round move far enough off course to not make a kill shot [after penetrating through glass]." He also believes that both the *trap* and *lead* methods have their place, dependent on the situation: "I do a little of both. Know them both and use them wisely."

McPhee was also instrumental in refining AVI or aerial vehicle interdiction techniques in both Iraq and Afghanistan. He comments: "We shot from every platform. Helos aren't that hard and modern war has shown the aerial interdiction is very effective. I did so many aerial missions I got an Air Medal. I found none of this hard."

McPhee would load the first five rounds in his SR–25K's magazine with armor-piercing bullets to penetrate the engine block or if necessary the windscreen glass. "VI taking targets from the air, [was] pretty simple really. Helos are a great way to get on top of the target quick. It's always a great way to get out of the area quick [too], leading to needing smaller numbers [of operators]. As a target is mobile, they are vunerable and this is the best time to take them."

Finally, McPhee was asked the million-dollar question: what was his opinion of *American Sniper*? "*American Sniper* is a series of beer-drinking tales. When told over beer [they] sound awesome. When fact checked [they] fall short on facts."

APPENDIX 2
REFERENCES

PRIMARY SOURCE MATERIAL

Mathew Coombes, former Australian Federal Police Specialist Response Group, Australia

John McPhee, former United States Army Special Operations Command 1st SFOD-D, USA

Nathan Vinson, former Australian Army Infantry Sniper School, Australia

SECONDARY SOURCE MATERIAL

Afong, Milo S., *Hunters: U.S. Snipers in the War on Terror* (New York: Berkley Caliber, 2010)

Cartwright, James, *Sniper in Helmand: Six Months on the Frontline* (Barnsley: Pen & Sword, 2011)

Cavallaro, Gina & Larsen, Matt, *Sniper: American Single-Shot Warriors in Iraq and Afghanistan* (Guildford: Lyons Press, 2010)

Coughlin, Gunnery Sergeant Jack, *Shock Factor: American Snipers in the War on Terror* (New York: St. Martin's Press, 2014)

Fury, Dalton, *Kill Bin Laden: A Delta Force Commander's Account of the Hunt for the World's Most Wanted Man* (New York: St. Martin's Press, 2008)

Gilbert, Adrian, *Stalk and Kill: The Sniper Experience* (London: Sidgwick & Jackson, 1997)

Golembesky, Michael, with Bruning, John R., *Level Zero Heroes: The Story of U.S. Marine Special Operations in Bala Murghab, Afghanistan* (New York: St. Martin's Press, 2014)

Halberstadt, Hans, *Trigger Men: Shadow Team, Spider-Man, The Magnificent Bastards, and the American Combat Sniper* (New York: St. Martin's Press, 2008)

Harnden, Toby, *Dead Men Risen: The Welsh Guards and the Real Story of Britain's War in Afghanistan* (London: Quercus, 2011)

Harrison, Sergeant Craig, *The Longest Kill: The Story of Maverick 41, One of the World's Greatest Snipers* (London: Sidgwick & Jackson, 2015)

Hogg, Ian V., *The World's Sniping Rifles* (London: Greenhill Books, 1998)

Irving, Nicholas, with Brozek, Gary, *The Reaper: Autobiography of One of the Deadliest Special Ops Snipers* (New York: St. Martin's Press, 2015)

Johnson, Dillard, and Tarr, James, *Carnivore: A Memoir by One of the Deadliest American Soldiers of All Time* (New York: William Morrow, 2013)

Kemp, Colonel Richard, & Hughes, Chris, *Attack State Red* (London: Michael Joseph, 2009)

Kyle, Chris, with McEwan, Scott and DeFelice, Jim, *American Sniper: The Autobiography of the Most Lethal Sniper in U.S. Military History* (New York: William Morrow, 2012)

Martin, Chris, with SOFREP.com, *Modern American Snipers: From the Legend to the Reaper – On the Battlefield with Special Operations Snipers* (New York: St. Martin's Press, 2014)

Mills, Sergeant Dan, *Sniper One: The Blistering True Story of a British Battle Group under Siege* (London: Michael Joseph, 2007)

Monty B., *A Sniper's Conflict: An Elite Sharpshooter's Thrilling Account of Hunting Insurgents in Afghanistan and Iraq* (New York: Skyhorse, 2014)

Plaster, Major John L., *The History of Sniping and Sharpshooting* (Boulder: Paladin Press, 2008)

Plaster, Major John L., *The Ultimate Sniper: An Advanced Training Manual for Military and Police Snipers* (Boulder: Paladin Press, 2006)

Scott, Jake, *Blood Clot: In Combat with the Patrols Platoon, 3 Para, Afghanistan, 2006* (Solihull: Helion, 2008)

Spicer, Mark, *Illustrated Manual of Sniper Skills* (Minneapolis: Zenith Press, 2006)

Tootal, Colonel Stuart, *Danger Close: Commanding 3 PARA in Afghanistan* (London: John Murray, 2009)

Wahlert, Glenn, & Linwood, Russell, *One Shot Kills: A History of Australian Army Sniping* (Canberra: Army History Unit, 2014)

NOTES

CHAPTER 1

1 "Overwatch" is a "tactical movement technique in which one element is positioned to support the movement of another element with immediate fire," according to the U.S. Army.

2 Lester Grau and Charles Q. Cutshaw, "Russian Snipers in the Mountains and Cities of Chechnya," *Infantry Magazine*, Volume 92 Number 2 (Summer 2002)

CHAPTER 2

1 Sergeant Eddie Waring, "My War: The Sniper," *BBC News* http://news.bbc.co.uk/2/hi/uk_news/2969255.stm

2 Jim Michaels, "U.S. Military Snipers are Changing Warfare," *USA Today* http://usatoday30.usatoday.com/news/military/story/2012-04-23/snipers-warfare-technology-training/54845142/1

3 Master Sergeant Duff E. McFadden, "Sniper Training Provides 'Combat Multiplier' for Iraqi Army," *DVIDS* https://www.dvidshub.net/news/47989/sniper-training-provides-combat-multiplier-iraqi-army#.VtOPnMtunIU

4 Charles222, "Combined Arms at the Platoon and Company Level," *The Firearms Blog* http://www.thefirearmblog.com/blog/2011/06/01/combined-arms-at-the-platoon-and-

company-level/

5 Navy SEAL snipers, for instance, have parachuted into the
ocean before later conducting a simultaneous shoot to eliminate
three Somali pirates holding a hostage, in one famous mission.
Canadian Army snipers have braved infiltration across almost
3,000-meter (3,280-yard) high, snow-capped Afghan mountains
before making a world record shot. British SAS snipers covertly
infiltrated into an apartment complex in Baghdad dressed as
locals to conduct an operation to stop a suicide bomber cell.

6 Gina Cavallaro and Matt Larsen, *Sniper: American Single-Shot
Warriors in Iraq and Afghanistan* (Guildford: Lyons Press, 2010), 50

7 Mark Spicer, *Illustrated Manual of Sniper Skills* (Minneapolis:
Zenith Press, 2006), 106

8 Sergeant 1st Class Shelman Spencer, "Sniper Competition Test
More Than Just Marksmanship," *DVIDS* https://www.dvidshub.
net/news/161524/sniper-competition-test-more-than-just-
marksmanship#.VtOmYstunIU

9 Staff Sergeant Coltin Heller, "GTA Provides Better Training, Say
Dutch Snipers," http://www.army.mil/article/134984

10 Roxana Tiron, "Canadian Army Snipers Gain From Afghanistan
Experience," *National Defense Magazine* http://www.
nationaldefensemagazine.org/archive/2004/January/Pages/
Canadian3665.aspx

11 Christian Marquardt, "Training Takes Special Operations
Snipers to New Heights in Europe," http://www.army.mil/
article/113878

12 Major General Walter Wojdakowski, "Adaptability – The Key To
Success In Mountain Operations," *Infantry Magazine*, Volume 97
Number 1 (January–February 2008)

13 1st Lt Mackenzie Eason, "1-24 IN Conducts Squad Designated
Marksman Training," http://www.army.mil/article/109921

14 Both units use the term *recce* rather than the more American *recon*

owing to their shared heritage with the British SAS and SBS.

15 Sergeant Craig Harrison, *The Longest Kill: The Story of Maverick 41, One of the World's Greatest Snipers* (London: Sidgwick & Jackson, 2015)

16 Cavallaro, *Sniper: American Single-Shot Warriors*, 51–52

17 http://www.combatstress.org.uk/medical-professionals/what-is-ptsd/

18 J. Peter Bradley, "An Exploratory Study on Sniper Well-Being," http://cradpdf.drdc-rddc.gc.ca/PDFS/unc102/p534015_A1b.pdf

19 Ibid.

20 Martin Bentham, "British Snipers Kill Four Iraqis Amid the Rubble of Basra," http://www.telegraph.co.uk/news/worldnews/northamerica/usa/1426504/British-snipers-kill-four-Iraqis-amid-the-rubble-of-Basra.html

CHAPTER 3

1 Cavallaro, *Sniper: American Single-Shot Warriors*, 14

2 Staff Sergeant Coltin Heller, "GTA Provides Better Training, Say Dutch Snipers," http://www.army.mil/article/134984

3 "Return of the Sniper: How Ancient Skills are Experiencing a Modern Renaissance in Afghanistan," http://www.independent.co.uk/news/world/asia/return-of-the-sniper-how-ancient-skills-are-experiencing-a-modern-renaissance-in-afghanistan-1727300.html

4 Carl Schulze, "ESTCON Snipers," *Combat & Survival Magazine*, February 2011

5 Glenn Wahlert and Russell Linwood, *One Shot Kills: A History of Australian Army Sniping* (Canberra: Army History Unit, 2014), 202

6 Carl Schulze, "Danish Snipers," *Combat & Survival Magazine*, March 2012

NOTES

7   Glenn Wahlert and Russell Linwood, *One Shot Kills*, 205

8   Greg Roberts, "Insurgency Sniping," *Combat & Survival Magazine*, March 2009

9   Carl Schulze, "Danish Snipers," *Combat & Survival Magazine*, March 2012

10 Colonel Richard Kemp and Chris Hughes, *Attack State Red* (London: Michael Joseph, 2009), 369.

11 Monty B, *A Sniper's Conflict: An Elite Sharpshooter's Thrilling Account of Hunting Insurgents in Afghanistan and Iraq* (New York: Skyhorse, 2014), 24

12 Toby Harnden, *Dead Men Risen: The Welsh Guards and the Real Story of Britain's War in Afghanistan* (London: Quercus, 2011), 434

13 "Sniper Kills Six Taliban with One Bullet," http://www.theguardian.com/uk-news/2014/apr/01/sniper-kills-five-taliban-with-one-bullet

14 "Marine Corps Cites Problems with Outdated Sniper Rifles," http://triblive.com/usworld/nation/8560879-74/marine-sniper-corps

15 Monty B, *A Sniper's Conflict*, 27

16 Colonel Stuart Tootal, *Danger Close: Commanding 3 PARA in Afghanistan* (London: John Murray, 2009), 121

17 Carl Schulze, "Troops In Contact!," *Combat & Survival Magazine*, October 2011

18 Keith Rogers, "Three-man Sniper Team Fends Off Taliban Onslaught during Operation *Bacha Strga*," *American Valor*, http://www.americanvalor.net/heroes/1934

19 James Cartwright, *Sniper in Helmand: Six Months on the Frontline* (Barnsley: Pen & Sword, 2011), 44

20 Toby Harnden, *Dead Men Risen*, 412

21 Glenn Wahlert and Russell Linwood, *One Shot Kills*, 196

22 Carl Schulze, "Chainsaw Snipers," *Combat & Survival Magazine*, June 2010

23 www.dtic.mil/get-tr-doc/pdf?AD=ADA512331

24 Dr Peter Pedersen, "The Falling Leaves of Tizak," *Wartime Magazine*, Issue 57 (January 2012), 14

25 Nicholas Irving with Gary Brozek, *The Reaper: Autobiography of one of the Deadliest Special Ops Snipers* (New York: St. Martin's Press, 2015), 21

CHAPTER 4

1 Martin Bentham, "British Snipers Kill Four Iraqis Amid the Rubble of Basra," http://www.telegraph.co.uk/news/worldnews/northamerica/usa/1426504/British-snipers-kill-four-Iraqis-amid-the-rubble-of-Basra.html

2 Sergeant Eddie Waring, "My War: The Sniper," *BBC News* http://news.bbc.co.uk/2/hi/uk_news/2969255.stm

3 Martin Bentham, "Snipers Aim at Enemy Morale," *The Age* http://www.theage.com.au/articles/2003/04/03/1048962880133.html

4 "Welsh Marksman Bends Shot to Kill Iraqi Rifleman," *Wales Online* http://www.walesonline.co.uk/news/wales-news/welsh-marksman-bends-shot-kill-2487044

5 Richard Lucas, "Spanish Legion," *Combat and Survival Magazine*, May 2012

6 C. J. Chivers, "Limited Success for U.S. Marine Snipers in Iraq," *New York Times* http://www.nytimes.com/2006/11/21/world/americas/21iht-web.1122sniper.3618580.html?pagewanted=all&_r=0

7 Matthew Cox, "Sniper's Skills in Demand in Iraq," http://www.snipercountry.com/Articles/SniperSkillsDemandIraq.asp

8 Jonathan Miltimore, "A Sniper's Tale of Survival in Ramadi," *Scout* http://www.scout.com/military/warrior/story/1506573-a-sniper-s-tale-of-survival-in-ramadi

NOTES

9 Dillard Johnson and James Tarr, *Carnivore: A Memoir by One of the Deadliest American Soldiers of All Time* (New York: William Morrow, 2013), 242

10 Major John L. Plaster, "Winning the Sniper War in Iraq," *American Rifleman* http://www.americanrifleman.org/articles/2010/1/14/winning-the-sniper-war-in-iraq/

11 Chris Hedges and Laila Al-Arian, "What Was This For?," *The Guardian* http://www.theguardian.com/world/2007/jul/13/usa.iraq2

12 Sergeant Dan Mills, *Sniper One: The Blistering True Story of a British Battle Group Under Siege* (London: Michael Joseph, 2007) 338

13 "Return of the Sniper: How Ancient Skills are Experiencing a Modern Renaissance in Afghanistan," http://www.independent.co.uk/news/world/asia/return-of-the-sniper-how-ancient-skills-are-experiencing-a-modern-renaissance-in-afghanistan-1727300.html

14 Simon Williams, "Airborne Sniper Squad Targets Iraqi Militia," http://www.raf.mod.uk/news/archive.cfm?storyid=5B9B5C07-1143-EC82-2E153B171A334FC4

15 Sergeant Eddie Waring, "My War: The Sniper," *BBC News* http://news.bbc.co.uk/2/hi/uk_news/2969255.stm

16 Cpl. Shawn C. Rhodes, "Marine Sniper Team Foils Roadside Bomb Attack," http://www.1stmardiv.marines.mil/News/NewsArticleDisplay/tabid/8585/Article/540780/marine-sniper-team-foils-roadside-bomb-attack.aspx

17 Jonathan Miltimore, "A Sniper's Tale of Survival in Ramadi," *Scout* http://www.scout.com/military/warrior/story/1506573-a-sniper-s-tale-of-survival-in-ramadi

18 Matthew Cox, "Sniper's Skills Keep Buddies Alive," *USA Today* http://usatoday30.usatoday.com/news/world/iraq/2003-12-26-sniper-usat_x.htm

19 Mark Spicer, *Illustrated Manual of Sniper Skills* (Minneapolis: Zenith Press, 2006)

20 Major John L. Plaster, *The History of Sniping and Sharpshooting* (Boulder: Paladin Press, 2008), 635

21 Cavallaro, *Sniper: American Single-Shot Warriors*, 70

22 Ibid.

23 Master Sergeant Duff E. McFadden, "Spartans Provide Iraqi Army with 'Combat multiplier,'" http://www.army.mil/article/37442/Spartans_provide_Iraqi_Army_with___039_combat_multiplier__039_/

24 Andrew Gilligan, "British Troops in Iraq Had to Let Attackers Go Free," *The Telegraph* http://www.telegraph.co.uk/news/uknews/defence/6639450/British-troops-in-Iraq-had-to-let-attackers-go-free.html

25 Tony Perry, "Marine Sergeant Wins Silver Star for Iraq Combat," http://www.leatherneck.com/forums/archive/index.php/t-20622.html

26 Cpl Paul W. Leicht, "Marine Sniper Credited with Longest Confirmed Kill in Iraq," *Marine Corps Times* http://www.freerepublic.com/focus/news/1313501/posts

CHAPTER 5

1 Greg Roberts, "Insurgency Sniping," *Combat & Survival Magazine*, February 2009

2 Cavallaro, *Sniper: American Single-Shot Warriors*, 59–60

3 Major John L. Plaster, "Winning the Sniper War in Iraq," *American Rifleman* http://www.americanrifleman.org/articles/2010/1/14/winning-the-sniper-war-in-iraq/

4 Mark Sixbey, "Sniper Kills Insurgent," *Marine Corps News* http://usmcronbo.tripod.com/id81.htm

5 Monte Morin, "Juba the Sniper Legend Haunting Troops in

Iraq," *Stars & Stripes* http://www.stripes.com/news/juba-the-sniper-legend-haunting-troops-in-iraq-1.63062

6   Cavallaro, *Sniper: American Single-Shot Warriors*, 61

7   Chris Kyle with Scott McEwan and Jim DeFelice, *American Sniper: The Autobiography of the Most Lethal Sniper in U.S. Military History* (New York: William Morrow, 2012), 139

8   Dillard Johnson, "Out Dueling a Master Iraqi Sniper," *Soldier of Fortune Magazine* https://www.sofmag.com/out-dueling-a-master-iraqi-sniper/

9   "Return of the Sniper: How Ancient Skills are Experiencing a Modern Renaissance in Afghanistan," http://www.independent.co.uk/news/world/asia/return-of-the-sniper-how-ancient-skills-are-experiencing-a-modern-renaissance-in-afghanistan-1727300.html

10 Dillard Johnson and James Tarr, *Carnivore: A Memoir by One of the Deadliest American Soldiers of All Time* (New York: William Morrow, 2013), 176

11 Colonel Richard Kemp and Chris Hughes, *Attack State Red* (London, Michael Joseph, 2009), 343

12 C. J. Chivers, "The Weakness of Taliban Marksmanship," *New York Times*, http://atwar.blogs.nytimes.com/2010/04/02/the-weakness-of-taliban-marksmanship/

13 Christopher Leake, "Taliban Snipers who Killed Ten British Soldiers Die in Air Strike," *Daily Mail* http://www.dailymail.co.uk/news/article-1305099/Taliban-snipers-killed-British-soldiers-die-air-strike.html

CHAPTER 6

1   Cavallaro, *Sniper: American Single-Shot Warriors*, 61

2   Mark Spicer, *Illustrated Manual of Sniper Skills*, 93

CHAPTER 7

1 Chris Kyle, *American Sniper*, 100

2 Sergeant Dan Mills, *Sniper One*, 84

3 Chris Kyle, *American Sniper*, p.103

4 Sergeant Dan Mills, *Sniper One*, 84

5 Sergeant Craig Harrison, *The Longest Kill*, 168

6 Dillard Johnson and James Tarr, *Carnivore*, 239

7 Chris Kyle, *American Sniper*, 102

8 James Cartwright, *Sniper in Helmand: Six Months on the Frontline* (Barnsley: Pen & Sword, 2011), 39

9 Colonel Richard Kemp and Chris Hughes, *Attack State Red*, 365

10 Lester Grau and Charles Q. Cutshaw, "Russian Snipers in the Mountains and Cities of Chechnya"

11 Toby Harnden, "Dead Men Risen: The Snipers' Story," *The Telegraph* http://www.telegraph.co.uk/culture/books/8376808/Dead-Men-Risen-The-snipers-story.html]

INDEX

INDEX